Pehr Kalm, Johann Reinhold Forster

Travels into North America

Containing its natural history. Vol. 3

Pehr Kalm, Johann Reinhold Forster

Travels into North America
Containing its natural history. Vol. 3

ISBN/EAN: 9783337344856

Printed in Europe, USA, Canada, Australia, Japan

Cover: Foto ©ninafisch / pixelio.de

More available books at **www.hansebooks.com**

TRAVELS
INTO
NORTH AMERICA;

CONTAINING

Its NATURAL HISTORY, AND
A circumstantial Account of its Plantations
and Agriculture in general,

WITH THE

CIVIL, ECCLESIASTICAL AND COMMERCIAL
STATE OF THE COUNTRY,

The MANNERS of the INHABITANTS, and several curious
and IMPORTANT REMARKS on various Subjects.

By PETER KALM,

Professor of Oeconomy in the University of *Aobo* in Swedish
Finland, and Member of the *Swedish* Royal Academy of
Sciences.

TRANSLATED INTO ENGLISH

By JOHN REINOLD FORSTER, F.A.S.

Enriched with a Map, several Cuts for the Illustration of
Natural History, and some additional Notes.

VOL. III.

LONDON:
Printed for the EDITOR;
And Sold by T. LOWNDES, in Fleet-street.
MDCCLXXI.

PREFACE

OF THE

EDITOR.

I COULD have left this volume without preface, was it not for some circumstances, which I am going to mention.

THE author of this account of *North-America* is a *Swede*, and therefore seems always to shew a peculiar way of thinking in regard to the *English* in general, and in regard to the first proprietors and inhabitants of *Philadelphia* in particular. The *French*, the natural enemies of the *English*, have, for upwards of a century, been the allies of the *Swedes*, who therefore are in general more fond of them than of the *English*. The external politeness of the *French* in

Canada fully captivated our author, prejudiced him in their favour, and alienated his mind, though unjustly, from the *English*. I have therefore now and then, in remarks, been obliged to do the *English* justice, especially when I saw the author carried away either by prejudice, misinformation, or ignorance. He passed almost all the winter, between 1748 and 1749, at *Raccoon*, and conversed there with his countrymen; when he came to *Philadelphia* he likewise was in the company of the *Swedes* settled there: these, no doubt, furnished him with many partial and disingenuous accounts of the *English*, and gave his mind that unfavourable bias which he so often displays in prejudice of a nation, now at the head of the enlightened world, in regard to every religious, moral, and social virtue. The author frequently seems to throw an illiberal reflection on the first proprietors of *Pensylvania*, and the quakers; though they got that province not by force, but by a charter from the *English* government, to whom the *Swedes* gave it up by virtue of a public treaty. Prompted by such false infi-

PREFACE. v

infinuations of his countrymen, he likewife enters very minutely into the circumftances of the *Swedes,* and often omits, or mifreprefents, more important points, relative to the legiflator and father of *Penfylvania, William Penn,* who gave that province exiftence, laws, and reputation. The accounts in the firft Volume, p. 32 and 33, 37, 42 and 46, feem to be founded on fuch mifreprefentations. A *philofopher* fhould examine fuch accounts, hear both parties, and emancipate himfelf from narrownefs of mind and prejudice.

THE author, however, often does juftice to the excellent conftitution of *Penfylvania,* as may be feen Vol. I. p. 58, 59. and likewife pag. 270, 271.

THE author fpeaks of *ftones attracting the moifture of the air;* fee Vol. I. p. 35; this is fomewhat unphilofophically expreffed. No ftone attracts the moifture of the air, unlefs impregnated with faline particles; however, when the ftones are colder than the atmofphere, they then condenfe the moifture of the air on their furface: the

porous

porous stones absorb it immediately, but those of a more solid texture, as marbles, &c. keep it on their surface till it evaporates.

PAGE 36. The author represents the *white cedar-wood* as almost entirely destroyed; though at present, above twenty years after his account, it is still used in *Pensylvania;* and quantities of it to be had, sufficient both for home consumption, and exportation to the *West-India* islands.

PAGE 48. The river *Delaware* is called one of the *greatest rivers in the world*; here, I suppose, the author forgot a great many its superiors.

FOR the tenor of the above remarks I am indebted to a worthy friend and benefactor.

To the Errata of the first Volume must be referred the following: page 117, note, line 5, *easible*, read, *feasible*. P. 247, line 3 and 4, *forty seven*, read, *seventy four*. P. 298, line 13, *Originals*, read, *Orignals*.

A.

PREFACE.

A word more I must add about the *American Fauna* and *Flora,* which I promised in my proposals. The author, who, as far as I know, is still living, has not yet finished this work; these three volumes contain all that he has hitherto published relative to *America;* the journal of a whole year's travelling, and especially his expedition to the *Iroquese,* and fort *Niagara,* are still to come; which, as soon as they appear, if Providence spares my life and health, and if my situation allows of it, I will translate into *English*; and there are some hopes of obtaining the original from the author. He likewise often promises, in the course of this work, to publish a great *Latin* work, concerning the animals and plants of *North-America,* as far as he went through it; which would certainly make the small catalogue I could make, useless. It is likewise probable that the description of the animal kingdom will fall to the share of an abler pen than mine.

I here take the opportunity of returning my humble thanks to my friends, who

PREFACE.

who have generously promoted this publication; as without this public manner of acknowledging their favours, I would think myself guilty of ingratitude, which, in my opinion, is one of the most detestable vices.

London,
Febr. the 15*th,* 1771.

PETER

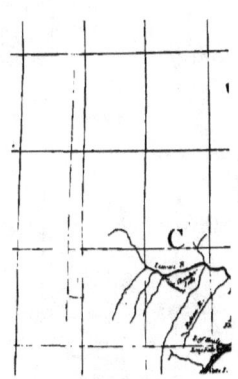

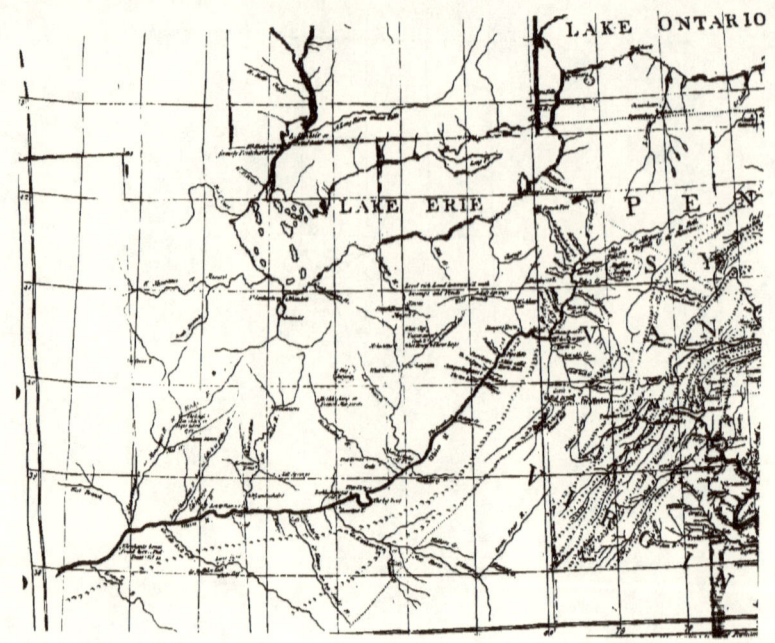

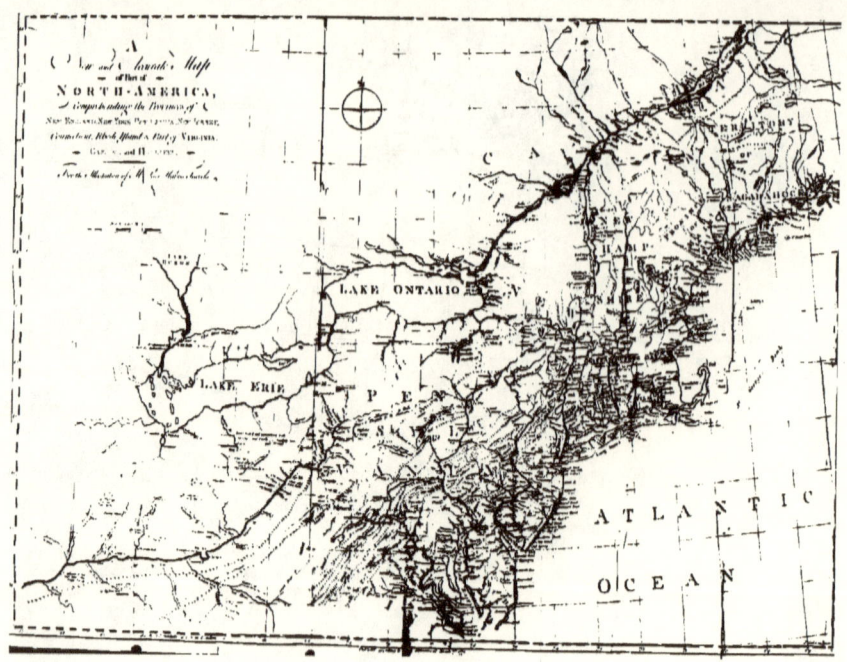

PETER KALM's TRAVELS.

July the 1st. 1749.

AT day break we got up, and rowed a good while before we got to the place where we left the true road. The country which we passed was the poorest and most disagreeable imaginable. We saw nothing but a row of amazing high mountains covered with woods, steep and dirty on their sides; so that we found it difficult to get to a dry place, in order to land and boil our dinner. In many places the ground, which was very smooth, was under water, and looked like the sides of our *Swedish* morasses which are intended to be drained; for this reason the *Dutch* in *Albany* call these parts the *Drowned Lands*.* Some of

* *De verdronkene landen.*

the mountains run from S. S. W. to N. N. E. and when they come to the river, they form perpendicular shores, and are full of stones of different magnitudes. The river runs for the distance of some miles together from south to north.

THE wind blew north all day, and made it very hard work for us to get forwards, though we all rowed as hard as we could, for our provisions were eaten to-day at breakfast. The river was frequently an *English* mile and more broad, then it became narrow again, and so on alternately; but upon the whole it kept a good breadth, and was surrounded on both sides by high mountains.

ABOUT six o'clock in the evening, we arrived at a point of land, about twelve *English* miles from Fort St. *Frederic*. Behind this point the river is converted into a spacious bay; and as the wind still kept blowing pretty strong from the north, it was impossible for us to get forwards, since we were extremely weak. We were therefore obliged to pass the night here, in spite of the *remonstrances* of our hungry stomachs.

IT is to be attributed to the peculiar grace of God towards us that we met the above mentioned *Frenchmen* on our journey,

and

and that they gave us leave to take one of their bark boats. It seldom happens once in three years, that the *French* go this road to *Albany*; for they commonly pass over the lake St. *Sacrement*, or, as the *English* call it, lake *George*, which is the nearer and better road, and every body wondered why they took this troublesome one. If we had not got their large strong boat, and been obliged to keep that which we had made, we would in all probability have been very ill off; for to venture upon the great bay during the least wind with so wretched a vessel, would have been a great piece of temerity, and we should have been in danger of being starved if we had waited for a calm. For being without fire-arms, and these deserts having but few quadrupeds, we must have subsisted upon frogs and snakes, which, (especially the latter) abound in these parts. I can never think of this journey, without reverently acknowledging the peculiar care and providence of the merciful Creator.

July the 2d. EARLY this morning we set out on our journey again, it being moon-shine and calm, and we feared lest the wind should change and become unfavourable to us if we stopped any longer. We all rowed as hard as possible, and happily arrived about eight in the morning at Fort St.

St. *Frederic*, which the *English* call *Crown Point*. Monsieur *Lusignan,* the governor, received us very politely. He was about fifty years old, well acquainted with polite literature, and had made several journies into this country, by which he had acquired an exact knowledge of several things relative to its state.

I was informed that during the whole of this summer, a continual drought had been here, and that they had not had any rain since last spring. The excessive heat had retarded the growth of plants; and on all dry hills the grass, and a vast number of plants, were quite dried up; the small trees, which grew near rocks, heated by the sun, had withered leaves, and the corn in the fields bore a very wretched aspect. The wheat had not yet eared, nor were the pease in blossoms. The ground was full of wide and deep cracks, in which the little snakes retired and hid themselves when pursued, as into an impregnable asylum.

The country hereabout, it is said, contains vast forests of firs of the white, black, and red kind, which had been formerly still more extensive. One of the chief reasons of their decrease are, the numerous fires which happen every year in the woods, through the carelessness of the *Indians,* who frequently

quently make great fires when they are hunting, which spread over the fir woods when every thing is dry.

GREAT efforts are made here for the advancement of *Natural History*, and there are few places in the world where such good regulations are made for this useful purpose, all which is chiefly owing to the care and zeal of a single person. From hence it appears, how well a useful science is received and set off, when the leading men of a country are its patrons. The governor of the fort, was pleased to shew me a long paper, which the then governor-general of *Canada*, the Marquis *la Galissonniere* had sent him. It was the same marquis, who some years after, as a *French* admiral, engaged the *English* fleet under admiral *Byng*, the consequence of which was the conquest of *Minorca*. In this writing, a number of trees and plants are mentioned, which grow in *North-America*, and deserve to be collected and cultivated on account of their useful qualities. Some of them are described, among which, is the *Polygala Senega*, or *Rattle Snake-root*; and with several of them the places where they grow are mentioned. It is further requested that all kinds of seeds and roots be gathered here; and, to assist such an undertaking, a method of preserving the gathered seeds and roots, is prescribed, so that they may grow, and be sent to *Paris*. Specimens of all kinds of minerals are required; and all the places in the *French* settlements are mentioned, where any useful or remarkable stone, earth, or ore has been found. There is likewise a manner of making observations and collections of curiosities in the animal kingdom. To these requests it is added, to enquire and get information, in every possible manner, to what purpose and in what manner the *Indians* employ certain plants and other productions of nature, as medicines, or in any other case. This useful paper was drawn up by order of the marquis *la Galissonniere*, by Mr. *Gaultier*, the royal physician at *Quebec*, and afterwards corrected and improved by the marquis's own hand. He had several copies made of it, which he sent to all the officers in the forts, and likewise to other learned men who travelled in the country. At the end of the writing is an injunction to the officers, to let the governor-general know, which of the common soldiers had used the greatest diligence in the discovery and collection of plants and other natural curiosities, that he might be able to promote them, when an opportunity occurred, to places adapted to their respective capacities, or to reward them in any other manner. I found that the people of distinction, in general here, had a much greater taste for natural history and other parts of literature, than in the *English* colonies, where it was every body's sole care and employment to scrape a fortune together, and where the sciences were held in universal contempt.* It was still complained

---

* It seems Mr. *Kalm* has forgotten his own assertions in the first volume. Dr. *Colden*, Dr. *Franklin*, and Mr. *Bartram*, have been the great promoters and investigators of nature in this country; and how would the inhabitants of *Old England* have gotten the fine collections of *North-American* trees, shrubs, and plants, which grow at present almost in every garden, and are as if it were naturalized in *Old England*, had they not been assisted by their friends, and by the curious in *North-America*. One need only cast an eye on Dr. *Linnæus's* new edition of his *Systema*, and the repeated mention of Dr. *Garden*, in order to be convinced that the *English* in *America* have contributed a greater share towards promoting natural history, than any nation under heaven, and certainly more than the *French*, though their learned men are often handsomely pensioned by their great *Monarque*: on the other hand the *English* study that branch of knowledge, from the sole motive of its utility, and the pleasure it affords to a thinking being, without any of those mercenary views, held forth to the learned of other countries. And as to the other parts of literature, the *English* in *America* are undoubtedly superior to the *French* in *Canada*, witness the many useful institutions, colleges, and schools founded in the *English* colonies in *North-America*, and so many very considerable libraries now erecting in this country, which contain such a choice of useful and curious books, as were very little known in *Canada*, before it fell into the hands of the *English*: not to mention the productions of original genius written by *Americans* born. F.

complained of here, that those who studied natural history, did not sufficiently enquire into the medicinal use of the plants of *Canada*.

The *French*, who are born in *France*, are said to enjoy a better health in *Canada* than in their native country, and to attain to a greater age, than the *French* born in *Canada*. I was likewise assured that the *European Frenchmen* can do more work, and perform more journies in winter, without prejudice to their health, than those born in this country. The intermitting fever which attacks the *Europeans* on their arrival in *Pensylvania*, and which as it were makes the climate familiar to them,* is not known here, and the people are as well after their arrival as before. The *English* have frequently observed, that those who are born in *America* of *European* parents, can never bear sea-voyages, and go to the different parts of *South America*, as well as those born in *Europe*. The *French* born in *Canada* have the same constitutions; and when any of them go to the *West-India* islands, such as *Martinique, Domingo*, &c. and make some stay there, they commonly fall sick and die soon after: those

* See Vol. I. p. 364.

who fall ill there seldom recover, unless they are brought back to *Canada*. On the contrary, those who go from *France* to those islands can more easily bear the climate, and attain a great age there, which I heard confirmed in many parts of *Canada*.

July the 5th. WHILST we were at dinner, we several times heard a repeated disagreeable outcry, at some distance from the fort, in the river *Woodcreek:* Mr. *Lusignan*, the governor, told us this cry was no good omen, because he could conclude from it that the *Indians*, whom we escaped near fort *Anne*, had completed their design of revenging the death of one of their brethren upon the *English*, and that their shouts shewed that they had killed an *Englishman*. As soon as I came to the window, I saw their boat, with a long pole at one end, on the extremity of which they had put a bloody skull. As soon as they were landed, we heard that they, being six in number, had continued their journey (from the place where we had marks of their passing the night), till they had got within the *English* boundaries, where they found a man and his son employed in mowing the corn. They crept on towards this man, and shot him dead upon the spot. This happened near the very village, where the *English*, two years before,

before, killed the brother of one of these *Indians*, who were then gone out to attack them. According to their custom they cut off the skull of the dead man, and took it with them, together with his clothes and his son, who was about nine years old. As soon as they came within a mile of fort St. *Frederic*, they put the skull on a pole, in the fore part of the boat, and shouted, as a sign of their success. They were dressed in shirts, as usual, but some of them had put on the dead man's clothes; one his coat, the other his breeches, another his hat, &c. Their faces were painted with vermillion, with which their shirts were marked across the shoulders. Most of them had great rings in their ears, which seemed to be a great inconvenience to them, as they were obliged to hold them when they leaped, or did any thing which required a violent motion. Some of them had girdles of the skins of *Rattle-snakes*, with the rattles on them; the son of the murdered man had nothing but his shirt, breeches and cap, and the *Indians* had marked his shoulders with red. When they got on shore, they took hold of the pole on which the skull was put, and danced and sung at the same time. Their view in taking the boy, was to carry him to
their

their habitations, to educate him inftead of their dead brother, and afterwards to marry him to one of their relations. Notwithftanding they had perpetrated this act of violence in time of peace, contrary to the command of the governor in *Montreal*, and to the advice of the governor of St. *Frederic*, yet the latter could not at prefent deny them provifions, and whatever they wanted for their journey, becaufe he did not think it advifeable to exafperate them; but when they came to *Montreal*, the governor called them to account for this action, and took the boy from them, whom he afterwards fent to his relations: Mr. *Lufignan* afked them, what they would have done to me and my companions, if they had met us in the defert? They replied, that as it was their chief intention to take their revenge on the *Englifhmen* in the village where their brother was killed, they would have let us alone; but it much depended on the humour they were in, juft at the time when we firft came to their fight. However, the commander and all the *Frenchmen* faid, that what had happened to me was infinitely fafer and better.

SOME years ago a fkeleton of an amazing great animal had been found in that part of *Canada*

Canada, where the *Illinois* live. One of the lieutenants in the fort assured me, that he had seen it. The *Indians*, who were there, had found it in a swamp. They were surprised at the sight of it, and when they were asked, what they thought it was? They answered that it must be the skeleton of the chief or father of all the beavers. It was of a prodigious bulk, and had thick white teeth, about ten inches long. It was looked upon as the skeleton of an elephant. The lieutenant assured me that the figure of the whole snout was yet to be seen, though it was half mouldered. He added, that he had not observed, that any of the bones were taken away, but thought the skeleton lay quite perfect there. I have heard people talk of this monstrous skeleton in several other parts of *Canada**.

BEARS are plentiful hereabouts, and they kept a young one, about three months old, at the fort. He had perfectly the same shape, and qualities, as our common bears in *Europe*, except the ears, which seemed to be longer in proportion, and the hairs which were stiffer; his colour was deep brown,

---

* THE country of the *Ilinois* is on the river *Ohio*, near the place where the *English* have found some bones, supposed to belong to elephants. See Vol. I. p. 135. in the note.

brown, almoſt black. He played and wreſtled every day with one of the dogs. A vaſt number of bear-ſkins are annually exported to *France* from *Canada*. The *Indians* prepare an oil from bear's greaſe, with which in ſummer they daub their face, hands, and all naked parts of their body, to ſecure them from the bite of the gnats. With this oil they likewiſe frequently ſmear the body, when they are exceſſively cold, tired with labour, hurt, and in other caſes. They believe it ſoftens the ſkin, and makes the body pliant, and is very ſerviceable to old age.

The common *Dandelion* (*Leontodon Taraxacum Linn.*) grows in abundance on the paſtures and roads between the fields, and was now in flower. In ſpring when the young leaves begin to come up, the *French* dig up the plants, take their roots [*], waſh them, cut them, and prepare them as a common ſallad; but they have a bitter taſte. It is not uſual here to make uſe of the leaves for eating.

July the 6th. The ſoldiers, which had been paid off after the war, had built houſes round the fort, on the grounds allotted

to

---

[*] In *France* the young blanched leaves, which ſcarce peep out of molehills, and have yet a yellow colour, are univerſally eaten as a ſallad, under the name of *Piſenlit*. F.

to them; but most of these habitations were no more than wretched cottages, no better than those in the most wreched places of *Sweden*; with that difference, however, that their inhabitants here were rarely oppressed by hunger, and could eat good and pure wheat bread. The huts which they had erected consisted of boards, standing perpendicularly close to each other. The roofs were of wood too. The crevices were stopped up with clay, to keep the room warm. The floor was commonly clay, or a black limestone, which is common here. The hearth was built of the same stone, except the place were the fire was to ly, which was made of grey sandstones, which for the greatest part consist of particles of quartz. In some hearths, the stones quite close to the fire-place were limestones; however, I was assured that there was no danger of fire, especially if the stones, which were most exposed to the heat, were of a large size. They had no glass in their windows.

July the 8th. THE *Galium tinctorium* is called *Tisavojaune rouge* by the *French* throughout all *Canada*, and abounds in the woods round this place, growing in a moist but fine soil. The roots of this plant are employed by the Indians in dying the quills of the *American* porcupines red, which they
put

put into several pieces of their work; and air, sun, or water seldom change this colour. The *French* women in *Canada* sometimes dye their clothes red with these roots, which are but small, like those of *Galium luteum*, or yellow bedstraw.

THE horses are left out of doors during the winter, and find their food in the woods, living upon nothing but dry plants, which are very abundant; however they do not fall off by this food, but look very fine and plump in spring.

July the 9th. THE skeleton of a whale was found some *French* miles from *Quebec*, and one *French* mile from the river *St. Laurence,* in a place where no flowing water comes to at present. This skeleton has been of a very considerable size, and the governor of the fort said, he had spoke with several people who had seen it.

July the 10th. THE boats which are here made use of, are of three kinds. 1. *Bark-boats,* made of the bark of trees, and of ribs of wood. 2. *Canoes,* consisting of a single piece of wood, hollowed out, which I have already described before [*]. They are here made of the white fir, and of different sizes. They are not brought

[*] See Vol. II.

forward by rowing, but by paddling; by which method not half the strength can be applied; which is made ufe of in rowing; and a fingle man might, I think, row as faft as two of them could paddle. 3. The third kind of boats are *Bateaux*. They are always made very large here, and employed for large cargoes. They are flat bottomed, and the bottom is made of the red, but more commonly of the white oak, which refifts better, when it runs againft a ftone, than other wood. The fides are made of the white fir, becaufe oak would make the *Bateau* too heavy. They make plenty of tar and pitch here.

The foldiery enjoy fuch advantages here, as they are not allowed in every part of the world. Thofe who formed the garrifon of this place, had a very plentiful allowance from their government. They get every day a pound and a half of wheat bread, which is almoft more than they can eat. They likewife get peafe, bacon, and falt meat in plenty. Sometimes they kill oxen and other cattle, the flefh of which is diftributed among the foldiers. All the officers kept cows, at the expence of the king, and the milk they gave was more than fufficient to fupply them. The foldiers had each a fmall garden without the fort, which they

they were allowed, to attend, and plant in it whatever they liked, and some of them had built summer-houses in them, and planted all kind of pot-herbs. The governor told me, that it was a general custom to allow the soldiers a spot of ground for kitchen-gardens, at such of the French forts hereabouts as were not situated near great towns, from whence they could be supplied with greens. In time of peace the soldiers have very little trouble with being upon guard at the fort; and as the lake close by is full of fish, and the woods abound with birds and animals, those amongst them who choose to be diligent, may live extremely well, and very grand in regard to food. Each soldier got a new coat every two years; but annually, a waistcoat, cap, hat, breeches, cravat, two pair of stockings, two pair of shoes, and as much wood as he had occasion for in winter. They likewise got five *sols** a piece every day; which is augmented to thirty sols when they have any particular labour for the king. When this is considered, it is not surprising to find the men are very fresh, well fed, strong and lively here. When a soldier falls sick he is brought to the hospital, where the king pro-

---

* A *sol* in *France* is about the value of one half penny sterling.

provides him with a bed, food, medicines, and people to take care of, and ferve him. When fome of them afked leave to be abfent for a day or two, to go abroad, it was generally granted them, if circumftances would permit, and they enjoyed as ufual their fhare of provifions and money, but were obliged to get fome of their comrades to mount the guard for them as often as it came to their turns, for which they gave them an equivalent. The governor and officers were duly honoured by the foldiers; however, the foldiers and officers often fpoke together as comrades, without any ceremonies, and with a very becoming freedom. The foldiers who are fent hither from *France*, commonly ferve till they are forty or fifty years old, after which they are difmiffed and allowed to fettle upon, and cultivate a piece of ground. But if they have agreed on their arrival to ferve no longer than a certain number of years, they are difmiffed at the expiration of their term. Thofe who are born here, commonly agree to ferve the crown during fix, eight, or ten years; after which they are difmiffed, and fet up for farmers in the country. The king prefents each difmiffed foldier with a piece of land, being commonly

monly 40 *arpens** long and but three broad, if the soil be of equal goodness throughout; but they get somewhat more, if it be a worse ground †. As soon as a soldier settles to cultivate such a piece of land, he is at first assisted by the king, who supplies himself, his wife and children, with provisions, during the three or four first years. The king likewise gives him a cow, and the most necessary instruments for agriculture. Some soldiers are sent to assist him in building a house, for which the king pays them. These are great helps to a poor man, who begins to keep house, and it seems that in a country where the troops are so highly distinguished by the royal favour, the king cannot be at a loss for soldiers. For the better cultivation and population of *Canada*, a plan has been proposed some years ago, for sending 300 men over from *France* every year, by which means the old

---

* An *Arpent* in *France* contains 100 *French* perches, and each of those 22 *French* feet; then the *French* foot being to the *English* as 1440 to 1352, an arpent is about 2346 *English* feet and 8 inches long. See *Ordonnances de Louis XIV. sur le fait des Eaux & Forêts.* Paris, 1687. p. 112. F.

† Mr. *Kalm* says, in his original, that the length of an *arpent* was so determined, that they reckoned 84 of them in a *French lieue* or league; but as this does by no means agree with the statute arpent of *France*, which by order of king *Lewis XIV*, was fixed at 2200 feet, *Paris* measure, (see the preceding note) we thought proper to leave it out of the text. F.

old foldiers may always be difmiffed, marry, and fettle in the country. The land which was allotted to the foldiers about this place, was very good, confifting throughout of a deep mould, mixed with clay.

July the 11th. THE harrows which they make ufe of here are made entirely of wood, and of a triangular form. The ploughs feemed to be lefs convenient. The wheels upon which the plough-beam is placed, are as thick as the wheels of a cart, and all the wood-work is fo clumfily made that it requires a horfe to draw the plough along a fmooth field.

ROCK-STONES of different forts lay fcattered on the fields. Some were from three to five feet high, and about three feet broad. They were pretty much alike in regard to the kind of the ftone, however, I obferved three different fpecies in them.

1. SOME confifted of a quartz, whofe colour refembled fugar candy, and which was mixed with a black fmall grained glimmer, a black horn-ftone, and a few minute grains of a brown fpar. The quartz was moft abundant in the mixture; the glimmer was likewife in great quantity, but the fpar was inconfiderable. The feveral kinds of ftones were well mixed, and though the eye could diftinguifh them, yet no inftrument

ment could separate them. The stone was very hard and compact, and the grains of quartz looked very fine.

2. Some pieces consisted of grey particles of quartz, black glimmer, and hornstone, together with a few particles of spar, which made a very close, hard, and compact mixture, only differing from the former in colour.

3. A few of the stones consisted of a mixture of white quartz and black glimmer, to which some red grains of quartz were added. The spar (quartz) was most predominant in this mixture, and the glimmer appeared in large flakes. This stone was not so well mixed as the former, and was by far not so hard and so compact, being easily pounded.

The mountains on which fort *St. Frederic* is built, as likewise those on which the above kinds of stone are found, consisted generally of a deep black lime-stone, lying in lamellæ as slates do, and it might be called a kind of slates, which can be turned into quicklime by fire[*]. This limestone is quite black in the inside, and, when broken, appears to be of an exceeding

---

[*] *Marmor schistosum*, Linn. Syst. III. p. 40. *Marmor unicolor nigrum.* Wal'. Min. pag. 61. n. 2. *Lime-slates, schistus calcareus,* Forst. Introd. to Min. p. 9. F.

ing fine texture. There are some grains of a dark spar scattered in it, which, together with some other inequalities, form veins in it. The strata which ly uppermost in the mountains consist of a grey limestone, which is seemingly no more than a variety of the preceding. The black limestone is constantly found filled with petrefactions of all kinds, and chiefly the following:

Pectinites, or petrefied *Ostreæ Pectines*. These petrefied shells were more abundant than any others that have been found here, and sometimes whole strata are met with, consisting merely of a quantity of shells of this sort, grown together. They are generally small, never exceeding an inch and a half in length. They are found in two different states of petrefaction; one shews always the impressions of the elevated and hollow surfaces of the shells, without any vestige of the shells themselves. In the other appears the real shell sticking in the stone, and by its light colour is easily distinguishable from the stone. Both these kinds are plentiful in the stone; however, the impressions are more in number than the real shells. Some of the shells are very elevated, especially in the middle, where they form as it were a hump; others again are

are depreſſed in the middle; but in moſt of them the outward ſurface is remarkably elevated. The furrows always run longitudinally, or from the top, diverging to the margin.

Petrefied Cornua Ammonis. Theſe are likewiſe frequently found, but not equal to the former in number: like the *pectinitæ*, they are found really petrefied, and in impreſſions; amongſt them were ſome petrefied ſnails. Some of theſe *Cornua Ammonis* were remarkably big, and I do not remember ſeeing their equals, for they meaſured above two feet in diameter.

DIFFERENT kinds of corals could be plainly ſeen in, and ſeparated from, the ſtone in which they lay. Some were white and ramoſe, or *Lithophytes*; others were ſtarry corals, or *Madrepores*; the latter were rather ſcarce.

I MUST give the name of *Stone-balls* to a kind of ſtones foreign to me, which are found in great plenty in ſome of the rockſtones. They were globular, one half of them projecting generally above the rock, and the other remaining in it. They conſiſt of nearly parallel fibres, which ariſe from the bottom as from a center, and ſpread over the ſurface of the ball and have a grey colour. The outſide of the balls is ſmooth, but

but has a number of small pores, which externally appear to be covered with a pale grey crust. They are from an inch to an inch and a half in diameter.

AMONGST some other kinds of sand, which are found on the shores of lake *Champlain*, two were very peculiar, and commonly lay in the same place; the one was black, and the other reddish brown, or granite coloured.

THE black sand always lies uppermost, consists of very fine grains, which, when examined by a microscope, appear to have a dark blue colour, like that of a smooth iron, not attacked by rust. Some grains are roundish, but most of them angular, with shining surfaces; and they sparkle when the sun shines. All the grains of this sand without exception are attracted by the magnet. Amongst these black or deep blue grains, they meet with a few grains of a red or garnet coloured sand, which is the same with the red sand which lies immediately under it, and which I shall now describe. This red or garnet coloured sand is very fine, but not so fine as the black sand. Its grains not only participate of the colour of garnets, but they are really nothing but pounded garnets. Some grains are round, others angulated; all shine and are

are femipellucid; but the magnet has no effect on them, and they do not fparkle fo much in funfhine. This red fand is feldom found very pure, it being commonly mixed with a white fand, confifting of particles of quartz. The black and red fand is not found in every part of the fhore, but only in a few places, in the order before mentioned. The uppermoft or black fand lay about a quarter of an inch deep; when it was carefully taken off, the fand under it became of a deeper red the deeper it lay, and its depth was commonly greater than that of the former. When this was carefully taken away, the white fand of quartz appeared mixed very much at top with the red fand, but growing purer the deeper it lay. This white fand was above four inches deep, had round grains, which made it entirely like a pearl fand. Below this was a pale grey angulated quartz fand. In fome places the garnet coloured fand lay uppermoft, and this grey angulated one immediately under it, without a grain of either the black or the white fand.

I cannot determine the origin of the black or fteel-coloured fand, for it was not known here whether there were iron mines in the neighbourhood or not. But I am rather inclined to believe they may be found
in

in these parts, as they are common in different parts of *Canada*, and as this sand is found on the shores of almost all the lakes, and rivers in *Canada*, though not in equal quantities. The red or garnet coloured sand has its origin hereabouts; for though the rocks near fort *St. Frederic* contained no garnets, yet there are stones of different sizes on the shores, quite different from the stones which form those rocks; these stones are very full of grains of garnets, and when pounded there is no perceptible difference between them and the red sand. In the more northerly parts of *Canada*, or below *Quebec*, the mountains themselves contain a great number of garnets. The garnet coloured sand is very common on the shores of the river *St. Laurence*. I shall leave out several observations which I made upon the minerals hereabouts, as uninteresting to most of my readers.

The *Apocynum androsæmifolium* grows in abundance on hills covered with trees, and is in full flower about this time; the *French* call it *Herbe à la puce*. When the stalk is cut or tore, a white milky juice comes out. The *French* attribute the same qualities to this plant, which the poison-tree, or *Rhus vernix*, has in the *English* colonies; that its poison is noxious to some per-

persons, and harmless to others. The milky juice, when spread upon the hands and body, has no bad effect on some persons; whereas others cannot come near it without being blistered. I saw a soldier whose hands were blistered all over, merely by plucking the plant, in order to shew it me; and it is said its exhalations affect some people, when they come within reach of them. It is generally allowed here, that the lactescent juice of this plant, when spread on any part of the human body not only swells the part, but frequently corrodes the skin; at least there are few examples of persons on whom it had no effect. As for my part, it has never hurt me, though in presence of several people I touched the plant, and rubbed my hands with the juice till they were white all over; and I have often rubbed the plant in my hands till it was quite crushed, without feeling the least inconvenience, or change on my hand. The cattle never touch this plant.

July the 12th. BURDOCK, or *Arctium Lappa*, grows in several places about the fort; and the governor told me, that its tender shoots are eaten in spring as raddishes, after the exterior peel is taken off.

THE *Sison Canadense* abounds in the woods

woods of all *North-America*. The *French* call it *cerfeuil fauvage*, and make use of it in spring, in green soups, like chervil. It is universally praised here as a wholesome, antiscorbutic plant, and as one of the best which can be had here in spring.

The *Asclepias Syriaca*, or, as the *French* call it, *le Cotonier*, grows abundant in the country, on the sides of hills which ly near rivers and other situations, as well in a dry and open place in the woods, as in a rich, loose soil. When the stalk is cut or broken it emits a lactescent juice, and for this reason the plant is reckoned in some degree poisonous. The *French* in *Canada* nevertheless use its tender shoots in spring, preparing them like asparagus; and the use of them is not attended with any bad consequences, as the slender shoots have not yet had time to suck up any thing poisonous. Its flowers are very odoriferous, and, when in season, they fill the woods with their fragrant exhalations, and make it agreeable to travel in them; especially in the evening. The *French* in *Canada* make a sugar of the flowers, which for that purpose are gathered in the morning, when they are covered all over with dew. This dew is expressed, and by boiling yields a very good brown, palatable sugar.

sugar. The pods of this plant when ripe contain a kind of wool, which encloses the seed, and resembles cotton, from whence the plant has got its *French* name. The poor collect it, and fill their beds, especially their children's, with it instead of feathers. This plant flowers in *Canada* at the end of *June* and beginning of *July*, and the seeds are ripe in the middle of *September*. The horses never eat of this plant.

July the 16th. THIS morning I crossed lake *Champlain* to the high mountain on its western side, in order to examine the plants and other curiosities there. From the top of the rocks, at a little distance from fort *St. Frederic*, a row of very high mountains appear on the western shore of lake *Champlain*, extending from south to north; and on the eastern side of this lake is another chain of high mountains, running in the same direction. Those on the eastern side are not close to the lake, being about ten or twelve miles from it; and the country between it and them is low and flat, and covered with woods, which likewise clothe the mountains, except in such places, as the fires, which destroy the forests here, have reached them and burnt them down. These mountains have generally steep sides, but sometimes they are found gradually sloping.

sloping. We crossed the lake in a canoe, which could only contain three persons, and as soon as we landed we walked from the shore to the top of the mountains. Their sides are very steep, and covered with a mould, and some great rock-stones lay on them. All the mountains are covered with trees; but in some places the forests have been destroyed by fire. After a great deal of trouble we reached the top of one of the mountains, which was covered with a dusty mould. It was none of the highest; and some of those which were at a greater distance were much higher, but we had no time to go to them; for the wind encreased, and our boat was but a little one. We found no curious plants, or any thing remarkable here.

When we returned to the shore we found the wind risen to such a height, that we did not venture to cross the lake in our boat, and for that reason I left the fellow to bring it back, as soon as the wind subsided, and walked round the bay, which was a walk of about seven *English* miles. I was followed by my servant, and for want of a road, we kept close to the shore where we passed over mountains and sharp stones; through thick forests and deep marshes, all which were known to be inhabited by

num-

numberless rattle-snakes, of which we happily saw none at all. The shore is very full of stones in some places, and covered with large angulated rock-stones, which are sometimes roundish, and their edges as it were worn off. Now and then we met with a small sandy spot, covered with grey, but chiefly with the fine red sand which I have before mentioned; and the black iron sand likewise occurred sometimes. We found stones of a red glimmer of a fine texture, on the mountains. Sometimes these mountains with the trees on them stood perpendicular with the waterside, but in some places the shore was marshy.

I saw a number of petrefied *Cornua Ammonis* in one place, near the shore, among a number of stones and rocks. The rocks consist of a grey limestone, which is a variety of the black one, and lies in strata, as that does. Some of them contain a number of petrefactions, with and without shells; and in one place we found prodigious large *Cornua Ammonis*, about twenty inches in breadth. In some places the water had wore off the stone, but could not have the same effect on the petrefactions, which lay elevated above, and in a manner glued on the stones.

THE

The mountains near the shore are amazingly high and large, consisting of a compact grey rock-stone, which does not ly in strata as the lime-stone, and the chief of whose constituent parts are a grey quartz, and a dark glimmer. This rock-stone reached down to the water, in places where the mountains stood close to the shore; but where they were at some distance from it, they were supplied by strata of grey and black lime-stone, which reached to the water side, and which I never have seen covered with the grey rocks.

The *Zizania aquatica* grows in mud, and in the most rapid parts of brooks, and is in full bloom about this time.

July the 17th. The distempers which rage among the *Indians* are *rheumatisms* and *pleurisies*, which arise from their being obliged frequently to ly in moist parts of the woods at night; from the sudden changes of heat and cold, to which the air is exposed here; and from their being frequently loaded with too great a quantity of strong liquor, in which case they commonly ly down naked in the open air, without any regard to the season, or the weather. These distempers, especially the pleurisies, are likewise very common among the *French* here; and the governor told me he had once had a very violent fit of the latter, and that Dr. *Sarrasin* had cured him in the following manner, which has been found to succeed best here. He gave him sudorifics, which were to operate between eight and ten hours; he was then bled, and the sudorifics repeated; he was bled again, and that effectually cured him.

Dr. *Sarrasin* was the royal physician at *Quebec*, and a correspondent of the royal academy of sciences at *Paris*. He was possessed of great knowledge in the practice of physic, anatomy, and other sciences, and very agreeable in his behaviour. He died at *Quebec*, of a malignant fever, which had been brought to that place by a ship, and with which he was infected at an hospital, where he visited the sick. He left a son, who likewise studied physic, and went to *France* to make himself more perfect in the practical part of it, but he died there.

The intermitting fevers sometimes come amongst the people here, and the venereal disease is common here. The *Indians* are likewise infected with it; and many of them have had it, and some still have it; but they likewise are perfectly possessed of the art of curing it. There are examples of *Frenchmen* and *Indians*, infected all over the body with this disease, who have been radically

dically and perfectly cured by the *Indians*, within five or six months. The *French* have not been able to find this remedy out; though they know that the *Indians* employ no mercury, but that their chief remedies are roots, which are unknown to the *French*. I have afterwards heard what these plants were, and given an account of them at large to the royal *Swedish* academy of sciences *.

WE are very well acquainted in *Sweden* with the pain caused by the *Tæniæ*, or a kind of worms. They are less abundant in the *British North-American* colonies; but in *Canada* they are very frequent. Some of these worms, which have been evacuated by a person, have been several yards long. It is not known, whether the *Indians* are afflicted with them, or not. No particular remedies against them are known here, and no one can give an account from whence they come, though the eating of some fruits contributes, as is conjectured, to create them.

July the 19th. FORT St. *Frederic* is a fortification, on the southern extremity of lake *Champlain*, situated on a neck of land, between that lake and the river, which arises from

---

* SEE the Memoirs of that Academy, for the year 1750. page 284.
  THE *Stillingia Sylvatica* is probably one of these roots. F.

from the union of the river *Woodcreek*, and lake St. *Sacrement*. The breadth of this river is here about a good musket shot. The *English* call this fortress *Crownpoint*, but its *French* name is derived from the *French* secretary of state, *Frederic Maurepas*, in whose hands the direction and management of the *French* court of admiralty was, at the time of the erection of this fort: for it is to be observed, that the government of *Canada* is subject to the court of admiralty in *France*, and the governor-general is always chosen out of that court. As most of the places in *Canada* bear the names of saints, custom has made it necessary to prefix the word *Saint* to the name of the fortress. The fort is built on a rock, consisting of black lime-slates, as afore said; it is nearly quadrangular, has high and thick walls, made of the same lime-stone, of which there is a quarry about half a mile from the fort. On the eastern part of the fort, is a high tower, which is proof against bombshells, provided with very thick and substantial walls, and well stored with cannon, from the bottom almost to the very top; and the governor lives in the tower. In the terre-plein of the fort is a well built little church, and houses of stone for the officers and soldiers. There are sharp rocks on on all sides towards the land, beyond a cannon-shot from the fort, but among them are some which are as high as the walls of the fort, and very near them.

The soil about fort St. *Frederic* is said to be very fertile, on both sides of the river; and before the last war a great many *French* families, especially old soldiers, have settled there; but the king obliged them to go into *Canada,* or to settle close to the fort, and to ly in it at night. A great number of them returned at this time, and it was thought that about forty or fifty families would go to settle here this autumn. Within one or two musket-shots to the east of the fort, is a wind-mill, built of stone with very thick walls, and most of the flour which is wanted to supply the fort is ground here. This wind-mill is so contrived, as to serve the purpose of a redoubt, and at the top of it are five or six small pieces of cannon. During the last war, there was a number of soldiers quartered in this mill, because they could from thence look a great way up the river, and observe whether the *English* boats approached; which could not be done from the fort itself, and which was a matter of great consequence, as the *English* might (if this guard had not been placed here) have gone in their little

boats

boats close under the western shore of the river, and then the hills would have prevented their being seen from the fort. Therefore the fort ought to have been built on the spot where the mill stands, and all those who come to see it, are immediately struck with the absurdity of its situation. If it had been erected in the place of the mill, it would have commanded the river, and prevented the approach of the enemy; and a small ditch cut through the loose limestone, from the river (which comes out of the lake St. *Sacrement*) to lake *Champlain*, would have surrounded the fort with flowing water, because it would have been situated on the extremity of the neck of land. In that case the fort would always have been sufficiently supplied with fresh water, and at a distance from the high rocks, which surround it in its present situation. We prepared to-day to leave this place, having waited during some days for the arrival of the yacht, which plies constantly all summer between the forts St. *John** and St. *Frederic*: during our stay here, we had received many favours. The governor of the fort, Mr. *Lusignan*, a man of learning and of great

---

* *Saint Jean.*

politeness, heaped obligations upon us, and treated us with as much civility as if we had been his relations. I had the honor of eating at his table during my stay here, and my servant was allowed to eat with his. We had our rooms, &c. to ourselves, and at our departure the governor supplied us with ample provisions for our journey to fort St. *John*. In short, he did us more favours than we could have expected from our own countrymen, and the officers were likewise particularly obliging to us.

About eleven o'clock in the morning we set out, with a fair wind. On both sides of the lake are high chains of mountains; with the difference which I have before observed, that on the eastern shore, is a low piece of ground covered with a forest, extending between twelve and eighteen *English* miles, after which the mountains begin; and the country behind them belongs to *New England*. This chain consists of high mountains, which are to be considered as the boundaries between the *French* and *English* possessions in these parts of *North America*. On the western shore of the lake, the mountains reach quite to the water side. The lake at first is but a *French* mile broad, but always encreases afterwards. The country is inhabited
within

within a *French* mile of the fort, but after that, it is covered with a thick foreſt. At the diſtance of about ten *French* miles from fort St. *Frederic*, the lake is four ſuch miles broad, and we perceive ſome iſlands in it. The captain of the yacht ſaid there were about ſixty iſlands in that lake, of which ſome were of a conſiderable ſize. He aſſured me that the lake was in moſt parts ſo deep, that a line of two hundred yards could not fathom it; and cloſe to the ſhore, where a chain of mountains generally runs acroſs the country, it frequently has a depth of eighty fathoms. Fourteen *French* miles from fort St. *Frederic* we ſaw four large iſlands in the lake, which is here about ſix *French* miles broad. This day the ſky was cloudy, and the clouds, which were very low, ſeemed to ſurround ſeveral high mountains, near the lake, with a fog; and from many mountains the fog roſe, as the ſmoke of a charcoal-kiln. Now and then we ſaw a little river which fell into the lake: the country behind the high mountains, on the weſtern ſide of the lake, is, as I am told, covered for many miles together with a tall foreſt, interſected by many rivers and brooks, with marſhes and ſmall lakes, and very fit to be inhabited. The ſhores are sometimes rocky, and sometimes sandy here. Towards night the mountains decreased gradually; the lake is very clear, and we observed neither rocks nor shallows in it. Late at night the wind abated, and we anchored close to the shore, and spent one night here.

July the 20th. THIS morning we proceeded with a fair wind. The place where we passed the night, was above half way to fort St. *John*; for the distance of that place from fort St. *Frederic*, across lake *Champlain* is computed to be forty-one *French* miles; that lake is here about six *English* miles in breadth. The mountains were now out of sight, and the country low, plain, and covered with trees. The shores were sandy, and the lake appeared now from four to six miles broad. It was really broader, but the islands made it appear narrower.

WE often saw *Indians* in bark-boats, close to the shore, which was however not inhabited; for the *Indians* came here only to catch sturgeons, wherewith this lake abounds, and which we often saw leaping up into the air. These *Indians* lead a very singular life: At one time of the year they live upon the small store of maize, beans, and melons, which they have planted; during another period, or about this time,

their

their food is fish, without bread or any other meat; and another season, they eat nothing but stags, roes, beavers, &c. which they shoot in the woods, and rivers. They, however, enjoy long life, perfect health, and are more able to undergo hardships than other people. They sing and dance, are joyful, and always content; and would not, for a great deal, exchange their manner of life for that which is preferred in *Europe*.

WHEN we were yet ten *French* miles from fort St. *John*, we saw some houses on the western side of the lake, in which the *French* had lived before the last war, and which they then abandoned, as it was by no means safe: they now returned to them again. These were the first houses and settlements which we saw after we had left those about fort St. *Frederic*.

THERE formerly was a wooden fort, or redoubt, on the eastern side of the lake, near the water-side; and the place where it stood was shewn me, which at present is quite overgrown with trees. The *French* built it to prevent the incursions of the *Indians*, over this lake; and I was assured that many *Frenchmen* had been slain in these places. At the same time they told me, that they reckon four women to one man

man in *Canada*, becaufe annually feveral *Frenchmen* are killed on their expeditions, which they undertake for the fake of trading with the *Indians*.

A WINDMILL, built of ftone, ftands on the eaft fide of the lake on a projecting piece of ground. Some *Frenchmen* have lived near it; but they left it when the war broke out, and are not yet come back to it. From this mill to fort St. *John* they reckon eight *French* miles. The *Englifh*, with their *Indians*, have burnt the houfes here feveral times, but the mill remained unhurt.

THE yacht which we went in to St. *John* was the firft that was built here, and employed on lake *Champlain*, for formerly they made ufe of *bateaux* to fend provifions over the lake. The Captain of the yacht was a *Frenchman*, born in this country; he had built it, and taken the foundings of the lake, in order to find out the true road, between fort St. *John* and fort St. *Frederic*. Oppofite the windmill the lake is about three fathoms deep, but it grows more and more fhallow, the nearer it comes to fort St. *John*.

WE now perceived houfes on the fhore again. The captain had otter-fkins in the cabin, which were perfectly the fame, in colour

colour and species, with the *European* ones. Otters are said to be very abundant in *Canada*.

Seal-skins are here made use of to cover boxes and trunks, and they often make portmantles of them in *Canada*. The common people had their tobacco-pouches made of the same skins. The seals here are entirely the same with the *Swedish* or *European* one, which are grey with black spots. They are said to be plentiful in the mouth of the river St. *Laurence*, below *Quebec*, and go up that river as far as its water is salt. They have not been found in any of the great lakes of *Canada*. The *French* call them *Loups marins*.*

The *French*, in their colonies, spend much more time in prayer and external worship, than the *English*, and *Dutch* settlers in the *British* colonies. The latter have neither morning nor evening prayer in their ships and yachts, and no difference is made between Sunday and other days. They never, or very seldom, say grace at dinner. On the contrary, the *French* here have prayers every morning and night on board their shipping, and on Sundays they pray more than commonly: they regularly say grace at their meals; and every one of

* Sea Wolves.

them

them says prayers in private as soon as he gets up. At fort St. *Frederic* all the soldiers assembled together for morning and evening prayers. The only fault was, that most of the prayers were read in *Latin*, which a great part of the people do not understand. Below the aforementioned wind-mill, the breadth of the lake is about a musket-shot, and it looks more like a river than a lake. The country on both sides is low and flat, and covered with woods. We saw at first a few scattered cottages along the shore; but a little further, the country is inhabited without interruption. The lake is here from six to ten foot deep, and forms several islands. During the whole course of this voyage, the situation of the lake was always directly from S. S. W. to N. N. E.

In some parts of *Canada* are great tracts of land belonging to single persons; from these lands, pieces, of forty *Arpens* long, and four wide, are allotted to each discharged soldier, who intends to settle here; but after his houshold is established, he is obliged to pay the owner of the lands six *French Francs* annually.

The lake was now so shallow in several places, that we were obliged to trace the way for the yacht, by sounding the depth
with

with branches of trees. In other places oppofite, it was fometimes two fathom deep.

In the evening, about fun fet, we arrived at fort St. *Jean*, or St. *John*, having had a continual change of rain, fun-fhine, wind, and calm, all the afternoon.

July the 21ft. St. *John* is a wooden fort, which the *French* built in 1748, on the weftern fhore of the mouth of lake *Champlain*, clofe to the water-fide. It was intended to cover the country round about it, which they were then going to people, and to ferve as a magazine for provifions and ammunition, which were ufually fent from *Montreal* to fort St. *Frederic*; becaufe they may go in yachts from hence to the laft mentioned place, which is impoffible lower down, as about two gunfhot further, there is a fhallow full of ftones, and very rapid water in the river, over which they can only pafs in *bateaux*, or flat veffels. Formerly fort *Chamblan*, which lies four *French* miles lower, was the magazine of provifions; but as they were forced firft to fend them hither in *bateaux*, and then from hence in yachts, and the road to fort *Chamblan* from *Montreal* being by land, and much round about, this fort was erected. It has a low fituation, and lies

in

in a sandy soil, and the country about it is likewise low, flat; and covered with woods. The fort is quadrangular, and includes the space of one *arpent* square. In each of the two corners which look towards the lake is a wooden building, four stories high, the lower part of which is of stone to the height of about a fathom and a half. In these buildings which are polyangular, are holes for cannon and lesser fire-arms. In each of the two other corners towards the country, is only a little wooden house, two stories high. These buildings are intended for the habitations of the soldiers, and for the better defence of the place; between these houses, there are poles, two fathoms and a half high, sharpened at the top, and driven into the ground close to one another. They are made of the *Thuya* tree, which is here reckoned the best wood for keeping from putrefaction, and is much preferable to fir in that point. Lower down the palisades were double, one row within the other. For the convenience of the soldiers, a broad elevated pavement, of more than two yards in height, is made in the inside of the fort all along the palisades, with a balustrade. On this pavement the soldiers stand and fire through the holes upon the enemy, without being exposed to
their

their fire. In the laſt year, 1748, two hundred men were in garriſon here; but at this time there were only a governor, a commiſſary, a baker, and ſix ſoldiers to take care of the fort and buildings, and to ſuperintend the proviſions which are carried to this place. The perſon who now commanded at the fort, was the Chevalier *de Gannes*, a very agreeable gentleman, and brother-in-law to Mr. *Luſignan*, the governor of fort St. *Frederic*. The ground about the fort, on both ſides of the water, is rich and has a very good ſoil; but it is ſtill without inhabitants, though it is talked of, that it ſhould get ſome as ſoon as poſſible.

THE *French* in all *Canada* call the gnats *Marangoins*, which name, it is ſaid, they have borrowed from the *Indians*. Theſe inſects are in ſuch prodigious numbers in the woods round fort St. *John*, that it would have been more properly called fort *de Marangoins*. The marſhes and the low ſituation of the country, together with the extent of the woods, contribute greatly to their multiplying ſo much; and when the woods will be cut down, the water drained, and the country cultivated, they probably will decreaſe in number, and vaniſh at laſt, as they have done in other places.

THE

The *Rattle Snake*, according to the unanimous accounts of the *French*, is never seen in this neighbourhood, nor further north near *Montreal* and *Quebec*; and the mountains which surround fort St. *Frederic*, are the most northerly part on this side, where they have been seen. Of all the snakes which are found in *Canada* to the north of these mountains, none is poisonous enough to do any great harm to a man; and all without exception run away when they see a man. My remarks on the nature and properties of the rattle-snake, I have communicated to the royal *Swedish* academy of sciences,* and thither I refer my readers.

July the 22d. This evening some people arrived with horses from *Prairie*, in order to fetch us. The governor had sent for them at my desire, because there were not yet any horses near fort St. *John*, the place being only a year old, and the people had not had time to settle near it. Those who led the horses, brought letters to the governor from the governor-general of *Canada*, the Marquis *la Galissoniere*, dated at *Quebec* the fifteenth of this month, and from the vice-governor of *Montreal*, the Baron

---

* See their Memoirs for the year 1752, p. 308, sect. 9.

de Longueil, dated the twenty-firſt of the
ſame month. They mentioned that I had
been particularly recommended by the
French court, and that the governor ſhould
ſupply me with every thing I wanted, and
forward my journey; and at the ſame time
the governor received two little caſks of
wine for me, which they thought would
relieve me on my journey. At night we
drank the kings of *France* and *Sweden*'s
health, under a ſalute from the cannon of
the fort, and the health of the governor-
general and others.

July the 23d. THIS morning we ſet
out on our journey to *Prairie*, from whence
we intended to proceed to *Montreal*; the
diſtance of *Prairie* from fort St *John*, by
land, is reckoned ſix *French* miles, and from
thence to *Montreal* two *lieues* (leagues)
and a half, by the river St. *Lawrence*. At
firſt we kept along the ſhore, ſo that
we had on our right the *Riviere de St. Jean*
(St. *John*'s river). This is the name of
the mouth of the lake *Champlain*, which
falls into the river St. *Lawrence*, and
is ſometimes called *Riviere de Champlain
(Champlain river.)* After we had travelled
about a *French* mile, we turned to the left
from the ſhore. The country was always
low, woody, and pretty wet, though it was

in the midſt of ſummer; ſo that we found it difficult to get forward. But it is to be obſerved that fort St. *John* was only built laſt ſummer, when this road was firſt made, and conſequently it could not yet have acquired a proper degree of ſolidity. Two hundred and ſixty men were three months at work, in making this road; for which they were fed at the expence of the government, and each received thirty ſols every day; and I was told that they would again reſume the work next autumn. The country hereabouts is low and woody, and of courſe the reſidence of millions of gnats and flies, which were very troubleſome to us. After we had gone about three *French* miles, we came out of the woods, and the ground ſeemed to have been formerly a marſh, which was now dried up. From hence we had a pretty good proſpect on all ſides. On our right hand at a great diſtance we ſaw two high mountains, riſing remarkably above the reſt; and they were not far from fort *Champlain*. We could likewiſe from hence ſee the high mountain which lies near *Montreal*; and our road went on nearly in a ſtraight line. Soon after, we got again upon wet and low grounds, and after that into a wood which conſiſted chiefly of

the

the fir with leaves which have a filvery underfide.* We found the foil which we paffed over to day, very fine and rich, and when the woods will be cleared and the ground cultivated, it will probably prove very fertile. There are no rocks, and hardly any ftones near the road.

ABOUT four *French* miles from fort St. *John*, the country makes quite another appearance. It is all cultivated, and a continual variety of fields with excellent wheat, peafe, and oats, prefented itfelf to our view; but we faw no other kinds of corn. The farms ftood fcattered, and each of them was furrounded by its corn fields, and meadows; the houfes are built of wood and very fmall. Inftead of mofs, which cannot be got here, they employ clay for ftopping up the crevices in the walls. The roofs are made very much floping, and covered with ftraw. The foil is good, flat, and divided by feveral rivulets; and only in a few places there are fome little hills. The profpect is very fine from this part of the road, and as far as I could fee the country, it was cultivated; all the fields were covered with corn, and they generally ufe fummer-wheat here. The ground is

* *Abies foliis fubtus argenteis.*

still very fertile, so that there is no occasion for leaving it ly as fallow. The forests are pretty much cleared, and it is to be feared that there will be a time, when wood will become very scarce. Such was the appearance of the country quite up to *Prairie*, and the river St. *Lawrence*, which last we had now always in sight; and, in a word this country was, in my opinion the finest of *North-America*, which I had hitherto seen.

About dinner-time we arrived at *Prairie*, which is situated on a little rising ground near the river St. *Lawrence*. We staid here this day, because I intended to visit the places in this neighbourhood, before I went on.

Prairie de la Magdelene is a small village on the eastern side of the river St. *Lawrence*, about two *French* miles and a half from *Montreal*, which place lies N. W. from hence, on the other side of the river. All the country round *Prairie* is quite flat, and has hardly any risings. On all sides are large corn-fields, meadows, and pastures. On the western side, the river St. *Lawrence* passes by, and has here a breadth of a *French* mile and a half, if not more. Most of the houses in *Prairie* are built of timber, with sloping wooden roofs, and the crevices in the

the walls are ftopped up with clay. There are fome little buildings of ftone, chiefly of the black lime-ftone, or of pieces of rock-ftone, in which latter the enchafement of the doors and windows was made of the black lime-ftone. In the midft of the village is a pretty church of ftone, with a fteeple at the weft end of it, furnifhed with bells. Before the door is a crofs, together with ladders, tongs, hammers, nails, &c. which are to reprefent all the inftruments made ufe of at the crucifixion of our Saviour, and perhaps many others befides them. The village is furrounded with palifades, from four yards to five high, put up formerly as a barrier againft the incurfions of the *Indians*. Without thefe palifades are feveral little kitchen and pleafure gardens, but very few fruit-trees in them. The rifing grounds along the river, are very inconfiderable here. In this place there was a prieft, and a captain, who affumed the name of governor. The cornfields round the place are extenfive, and fown with fummer-wheat; but rye, barley and maize are never feen. To the fouthweft of this place is a great fall in the river St. *Lawrence*, and the noife which it caufes, may be plainly heard here. When the water in fpring encreafes in the river, on account of the ice which then begins to diffolve, it fometimes happens to rife fo high as to overflow a great part of the fields, and, inftead of fertilizing them as the river *Nile* fertilizes the *Egyptian* fields by its inundations, it does them much damage, by carrying a number of graffes and plants on them, the feeds of which fpread the worft kind of weeds, and ruin the fields. Thefe inundations oblige the people to take their cattle a great way off, becaufe the water covers a great tract of land; but happily it never ftays on it above two or three days. The caufe of thefe inundations is generally owing to the ftopping of ice in fome part of the river.

The *Zizania aquatica*, or *Folle Avoine* grows plentiful in the rivulet, or brook, which flows fomewhat below *Prairie*.

July the 24th. This morning I went from *Prairie* in a bateau to *Montreal*, upon the river St. *Lawrence*. The river is very rapid, but not very deep near *Prairie*, fo that the yacht cannot go higher than *Montreal*, except in fpring with the high water, when they can come up to *Prairie*, but no further. The town of *Montreal* may be feen at *Prairie*, and all the way down to it. On our arrival, there we found a crowd of people at that gate of the town, where we

were

were to pafs through. They were very defirous of feeing us, becaufe they were informed that fome *Swedes* were to come to town; people of whom they had heard fomething, but whom they had never feen; and we were affured by every body, that we were the firft *Swedes* that ever came to *Montreal*. As foon as we were landed, the governor of the town fent a captain to me, who defired I would follow him to the governor's houfe, where he introduced me to him. The Baron *Longueuil* was as yet vice-governor, but he daily expected his promotion from *France*. He received me more civilly and generoufly than I can well defcribe, and fhewed me letters from the governor-general at *Quebec*, the Marquis *de la Galiffoniere*, which mentioned that he had received orders from the *French* court to fupply me with whatever I fhould want, as I was to travel in this country at the expence of his moft Chriftian majefty. In fhort governor *Longueuil* loaded me with greater favours than I could expect or even imagine, both during my prefent ftay and on my return from *Quebec*.

The difference between the manners and cuftoms of the *French* in *Montreal* and *Canada*, and thofe of the *Englifh* in the *American* colonies, is as great as that between

tween the manners of those two nations in
Europe. The women in general are handsome here ; they are well bred, and virtuous with an innocent and becoming freedom. They dress out very fine on Sundays;
and though on the other days they do not
take much pains with other parts of their
dress, yet they are very fond of adorning
their heads, the hair of which is always
curled and powdered, and ornamented with
glittering bodkins and aigrettes. Every
day but Sunday, they wear a little neat
jacket, and a short petticoat which hardly
reaches half the leg, and in this particular
they seem to imitate the *Indian* women.
The heels of their shoes are high, and very
narrow, and it is surprizing how they walk
on them. In their knowledge of œconomy,
they greatly surpass the *English* women in
the plantations, who indeed have taken the
liberty of throwing all the burthen of housekeeping upon their husbands, and sit in their
chairs all day with folded arms. * The
women in *Canada* on the contrary do not
spare themselves, especially among the common

---

* It seems, that for the future, the fair sex in the *English* colonies in *North-America*, will no longer deserve the reproaches Mr. Kalm stigmatizes them with repeatedly, since it is generally reported, that the ladies of late have vied one with another, in providing their families with linen, stockings, and home-spun coath of their own making, and that a general spirit of industry prevails among them at this present time. F.

mon people, where they are always in the fields, meadows, stables, &c. and do not dislike any work whatsoever. However, they seem rather remiss in regard to the cleaning of the utensils, and apartments; for sometimes the floors, both in the town and country, were hardly cleaned once in six months, which is a disagreeable sight to one who comes from amongst the *Dutch* and *English*, where the constant scouring and scrubbing of the floors, is reckoned as important as the exercise of religion itself. To prevent the thick dust, which is thus left on the floor, from being noxious to the health, the women wet it several times a day, which renders it more consistent; repeating the aspersion as often as the dust is dry and rises again. Upon the whole, however, they are not averse to the taking a part in all the business of housekeeping; and I have with pleasure seen the daughters of the better sort of people, and of the governor himself, not too finely dressed, and going into kitchens and cellars, to look that every thing be done as it ought.

THE men are extremely civil, and take their hats off to every person indifferently whom they meet in the streets. It is customary to return a visit the day after you have received one; though one should have some scores to pay in one day.

I HAVE

I HAVE been told by some among the *French*, who had gone a beaver-hunting with the *Indians* to the northern parts of *Canada*, that the animals, whose skins they endeavour to get, and which are there in great plenty, are beavers, wild cats, or lynxs, and martens. These animals are the more valued, the further they are caught to the north, for their skins have better hair, and look better than those which are taken more southward, and they became gradually better or worse, the more they are northward or southward.

White Patridges * is the name which the *French* in *Canada* give to a kind of birds, abounding during winter near *Hudson's Bay*, and which are undoubtedly our *Ptarmigans*, or *Snow-hens* (*Tetrao Lagopus*). They are very plentiful at the time of a great frost, and when a considerable quantity of snow happens to fall. They are described to me as having rough white feet, and being white all over, except three or four black feathers in the tail; and they are reckoned very fine eating. From *Edward*'s Natural History of Birds (pag. 72.) it appears, that the ptarmigans are common about *Hudson's Bay* †.

Hares

---

* Perdrix blanches.
† See Br. Zool. Suppl. plate XIII. f. 1. F.

Hares are likewife faid to be plentiful near *Hudfon's Bay*, and they are abundant even in *Canada*, where I have often feen, and found them perfectly correfponding with our *Swedifh* hares. In fummer they have a brownifh grey, and in winter a fnowy white colour, as with us *.

MECHANICS, fuch as architecture, cabinet-work, turning, and the like, were not yet fo forward here as they ought to be; and the *Englifh*, in that particular, out do the *French*. The chief caufe of this is, that fcarce any other people than difmiffed foldiers come to fettle here, who have not had any opportunity of learning a mechanical trade, but have fometimes accidentally, and through neceffity been obliged to it. There are however fome, who have a good notion of mechanics, and I faw a perfon here, who made very good clocks, and watches, though he had had but very little inftruction.

July the 27th. THE common houfe-flies have but been obferved in this country about one hundred and fifty years ago, as I have been affured by feveral perfons in this town, and in *Quebec*. All the *Indians* affert the fame thing, and are of opinion that the com-

---

* See a figure of this hare in its white ftate, in the Suppl. to *Br. Zool.* plate XLVII. f. 1. F.

common flies firſt came over here, with the *Europeans* and their ſhips, which were ſtranded on this coaſt. I ſhall not diſpute this; however, I know, that whilſt I was in the deſarts between *Saratoga* and *Crown-point*, or fort St. *Frederic*, and ſat down to reſt or to eat, a number of our common flies always came, and ſettled on me. It is therefore dubious, whether they have not been longer in *America* than the term above mentioned, or whether they have been imported from *Europe* On the other hand, it may be urged that the flies were left in thoſe deſarts at the time when fort *Anne* was yet in a good condition, and when the *Engliſh* often travelled there and back again; not to mention that ſeveral *Europeans*, both before and after that time, had travelled through thoſe places, and carried the flies with them, which were attracted by their proviſions.

Wild Cattle are abundant in the ſouthern parts of *Canada*, and have been there ſince times immemorial. They are plentiful in thoſe parts, particularly where the *Illinois Indians* live, which are nearly in the ſame latitude with *Philadelphia*; but further to the north they are ſeldom obſerved. I ſaw the ſkin of a wild ox to-day; it was as big as one of the largeſt ox hides in *Europe*,

rope, but had better hair. The hair is dark brown, like that on a brown bear-skin. That which is close to the skin, is as soft as wool. This hide was not very thick; and in general they do not reckon them so valuable as bear-skins in *France*. In winter they are spread on the floors, to keep the feet warm. Some of these wild cattle, as I am told, have a long and fine wool, as good, if not better, than sheep wool. They make stockings, cloth, gloves, and other pieces of worsted work of it, which look as well as if they were made of the best sheep wool; and the *Indians* employ it for several uses. The flesh equals the best beef in goodness and fatness. Sometimes the hides are thick, and may be made use of as cow-hides are in *Europe*. The wild cattle in general are said to be stronger and bigger, than *European* cattle, and of a brown red colour. Their horns are but short, though very thick close to the head. These and several other quali-ties, which they have in common with, and in greater perfection than the tame cattle, have induced some to endeavour to tame them; by which means they would obtain the advantages arising from their goodness of hair, and, on account of their great strength, be able to employ them

suc-

successfully in agriculture. With this view
some have repeatedly got young wild calves,
and brought them up in *Quebec*, and other
places, among the tame cattle; but they
commonly died in three or four years time;
and though they have seen people every day,
yet they have always retained a natural ferocity. They have constantly been very shy,
pricked up their ears at the sight of a
man, and trembled, or run about; so
that the art of taming them has not hitherto been found out. Some have been
of opinion, that these cattle cannot well
bear the cold; as they never go north of
the place I mentioned, though the summers
be very hot, even in those northern parts.
They think that, when the country about
the *Illinois* will be better peopled, it will be
more easy to tame these cattle, and that
afterwards they might more easily be used
to the northerly climates [*]. The *Indians*
and *French* in *Canada*, make use of the
horns of these creatures to put gun-powder
in. I have briefly mentioned the wild cattle in the former parts of this journey [†].

THE

---

[*] But by this means they would loose that superiority, which in their wild state they have over the tame cattle; as all the progenies of tamed animals degenerate from the excellence of their wild and free ancestors. F.

[†] See Vol. I. p. 207.

THE peace, which was concluded between *France* and *England,* was proclaimed this day. The foldiers were under arms; the artillery on the walls was fired off, and fome falutes were given by the fmall fire-arms. All night fome fireworks were exhibited, and the whole town was illuminated. All the ftreets were crowded with people, till late at night. The governor invited me to fupper, and to partake of the joy of the inhabitants. There were prefent a number of officers, and perfons of diftinction; and the feftival concluded with the greateft joy.

July the 28th. THIS morning I accompanied the governor, baron *Longueuil,* and his family, to a little ifland called *Magdelene,* which is his own property. It lies in the river St. *Lawrence,* directly oppofite to the town, on the eaftern fide. The governor had here a very neat houfe, though it was not very large, a fine extenfive garden, and a court-yard. The river paffes between the town and this ifland, and is very rapid. Near the town it is deep enough for yachts; but towards the ifland it grows more fhallow, fo that they are obliged to pufh the boats forwards with poles. There was a mill on the ifland, turned by the mere force of the ftream, without an additional mill-dam.

THE

The smooth sumach, or *Rhus giabra*, grows in great plenty here. I have no where seen it so tall as in this place, where it had sometimes the height of eight yards, and a proportionable thickness.

Sassafras is planted here; for it is never found wild in these parts, fort *Anne* being the most northerly place where I have found it wild. Those shrubs which were on the island, had been planted many years ago; however, they were but small shrubs, from two to three feet high, and scarce so much. The reason is, because the stem is killed every winter, almost down to the very root, and must produce new shoots every spring, as I have found from my own observations here; and so it appeared to be near the forts *Anne*, *Nicholson*, and *Oswego*. It will therefore be in vain to attempt to plant sassafras in a very cold climate.

The red Mulberry-trees *(Morus rubra, Linn.)* are likewise planted here. I saw four or five of them about five yards high, which the governor told me, had been twenty years in this place, and were brought from more southerly parts, since they do not grow wild near *Montreal*. The most northerly place, where I have found it growing spontaneously, is about twenty *English* miles north of *Albany*, as I have been

been affured by the country people, who live in that place, and who at the fame time informed me, that it was very fcarce in the woods. When I came to *Saratoga*, I enquired whether any of thefe mulberry-trees had been found in that neighbourhood? but every body told me, that they were never feen in thofe parts, but that the before mentioned place, twenty miles above *Albany*, is the moft northern one where they grow. Thofe mulberry-trees, which were planted on this ifland, fucceed very well, though they are placed in a poor foil. Their foliage is large and thick, but they did not bear any fruits this year. However, I was informed that they can bear a confiderable degree of cold.

The *Waterbeech* was planted here in a fhady place, and was grown to a great height. All the *French* hereabouts call it *Cotonier* *. It is never found wild near the river St. *Lawrence*; nor north of fort St. *Frederic*, where it is now very fcarce.

The red Cedar is called *Cedre rouge* by the *French*, and it was likewife planted in the governor's garden, whither it had been brought from more fouthern parts, for it is not to be met with in the forefts hereabouts.

---

* *Cotton-tree*. Mr. *Kalm* mentions before, that this name is given to the *Afclepias Syriaca*. See Vol. III. p. 28. F.

abouts. However, it came on very well here.

About half an hour after seven in the evening we left this pleasant island, and an hour after our return the baron *de Longueuil* received two agreeable pieces of news at once. The first was, that his son, who had been two years in *France*, was returned; and the second, that he had brought with him the royal patents for his father, by which he was appointed governor of *Montreal*, and the country belonging to it.

They make use of fans here, which are made of the tails of the wild turkeys. As soon as the birds are shot, their tails are spread like fans, and dried, by which means they keep their figure. The ladies and the men of distinction in town wear these fans, when they walk in the streets, during the intenseness of the heat.

All the grass on the meadows round *Montreal*, consists chiefly of a species of *Meadow-grass*, or the *Poa capillaris, Linn.* * This is a very slender grass, which grows very close, and succeeds even on the driest hills. It is however not rich in foliage; and the slender stalk is chiefly used for hay.
We

---

* Mr. *Kalm* describes it thus: *Poa culmo subcompresso, panicula tenuissima, spiculis trifloris minimis, flosculis basi pubescentibus.*

We have numerous kinds of grasses in *Sweden*, which make infinitely finer meadows than this.

July the 30th. THE *wild Plumb-trees* grow in great abundance on the hills, along the rivulets about the town. They were so loaded with fruit, that the boughs were quite bent downwards by the weight. The fruit was not yet ripe, but when it comes to that perfection, it has a red colour and a fine taste, and preserves are sometimes made of it.

Black Currants (*Ribes nigrum, Linn.*) are plentiful in the same places, and its berries were ripe at this time. They are very small, and not by far so agreeable as those in *Sweden*.

Parsneps grow in great abundance on the rising banks of rivers, along the corn-fields, and in other places. This led me to think, that they were original natives of *America*, and not first brought over by the *Europeans*. But on my journey into the country of the *Iroquois*, where no *European* ever had a settlement, I never once saw it, though the soil was excellent; and from hence it appears plain enough, that it was transported hither from *Europe*, and is not originally an *American* plant; and therefore it is in vain sought for in any part of this continent,

tinent, except among the *European* settlements,

August the 1st. THE governor-general of *Canada* commonly resides at *Quebec*; but he frequently goes to *Montreal*, and generally spends the winter there. In summer he chiefly resides at *Quebec*, on account of the king's ships, which arrive there during that season, and bring him letters, which he must answer; besides other business which comes in about that time. During his residence in *Montreal* he lives in the castle, as it is called, which is a large house of stone, built by governor-general *Vaudreuil*, and still belonging to his family, who hire it to the king. The governor-general *de la Galissoniere* is said to like *Montreal* better than *Quebec*, and indeed the situation of the former is by far the more agreeable one.

THEY have in *Canada* scarce any other but paper-currency. I hardly ever saw any coin, except *French* sols, consisting of brass, with a very small mixture of silver; they were quite thin by constant circulation, and were valued at a sol and a half. The bills are not printed, but written. Their origin is as follows. The *French* king having found it very dangerous to send money

for

for the pay of the troops, and other purposes, over to *Canada*, on account of privateers, shipwrecks, and other accidents; he ordered that instead of it the intendant, or king's steward, at *Quebec*, or the commissary at *Montreal*, is to write bills for the value of the sums which are due to the troops, and which he distributes to each soldier. On these bills is inscribed, that they bear the value of such or such a sum, till next *October*; and they are signed by the intendant, or the commissary; and in the interval they bear the value of money. In the month of *October*, at a certain stated time, every one brings the bills in his possession to the intendant at *Quebec*, or the commissary at *Montreal*, who exchanges them for bills of exchange upon *France*, which are paid there in lawful money, at the king's exchequer, as soon as they are presented. If the money is not yet wanted, the bill may be kept till next *October*, when it may be exchanged by one of those gentlemen, for a bill upon *France*. The paper money can only be delivered in *October*, and exchanged for bills upon *France*. They are of different values, and some do not exceed a *livre*, and perhaps some are still less. Towards autumn when the merchants ships come in from *France*, the merchants endeavour

deavour to get as many bills as they can, and change them for bills upon the *French* treasury. These bills are partly printed, spaces being left for the name, sum, &c. But the first bill, or paper currency is all wrote, and is therefore subject to be counterfeited, which has sometimes been done; but the great punishments, which have been inflicted upon the authors of these forged bills, and which generally are capital, have deterred people from attempting it again; so that examples of this kind are very scarce at present. As there is a great want of small coin here, the buyers, or sellers, were frequently obliged to suffer a small loss, and could pay no intermediate prices between one livre and two *.

They commonly give one hundred and fifty livres a year to a faithful and diligent footman, and to a maid-servant of the same character one hundred livres. A journeymen to an artist gets three or four livres a day, and a common labouring man gets thirty or forty sols a day. The scarcity of labouring people occasions the wages to be so high; for almost every body finds
it

---

* The *sol* is the lowest coin in *Canada*, and is about the value of a penny in the *English* colonies. A *livre*, or *franc*, (for they are both the same) contains twenty sols; and three livres, or francs, make an *ecu*, or crown.

it so easy to set up as a farmer in this uncultivated country, where he can live well, and at a small expence, that he does not care to serve and work for others.

Montreal is the second town in *Canada*, in regard to size and wealth; but it is the first on account of its fine situation, and mild climate. Somewhat above the town, the river St. *Lawrence* divides into several branches, and by that means forms several islands, among which the isle of *Montreal* is the greatest. It is ten *French* miles long, and near four broad, in its broadest part. The town of *Montreal* is built on the eastern side of the island, and close to one of the most considerable branches of the river St. *Lawrence*; and thus it receives a very pleasant, and advantageous situation. The town has a quadrangular form, or rather it is a rectangular parallelogram, the long and eastern side of which extends along the great branch of the river. On the other side it is surrounded with excellent corn-fields, charming meadows, and delightful woods. It has got the name of *Montreal* from a great mountain, about half a mile westwards of the town, and lifting its head far above the woods. Monf. *Cartier*, one of the first *Frenchmen* who surveyed *Canada* more accurately, called this
moun-

mountain so, on his arrival in this island, in the year 1535, when he visited the mountain, and the *Indian* town *Hoshelaga* near it. The priests who, according to the Roman catholic way, would call every place in this country after some saint or other, called *Montreal, Ville Marie*, but they have not been able to make this name general, for it has always kept its first name. It is pretty well fortified, and surrounded with a high and thick wall. On the east side it has the river St. *Lawrence*, and on all the other sides a deep ditch filled with water, which secures the inhabitants against all danger from the sudden incursions of the enemy's troops. However, it cannot long stand a regular siege, because it requires a great garrison, on account of its extent; and because it consists chiefly of wooden houses. Here are several churches, of which I shall only mention that belonging to the friars of the order of St. *Sulpitius*, that of the Jesuits, that of the *Franciscan* friars, that belonging to the nunnery, and that of the hospital; of which the first is however by far the finest, both in regard to its outward and inward ornaments, not only in this place, but in all *Canada*. The priests of the seminary of St. *Sulpitius* have a fine large house, where

they

they live together. The college of the *Francifcan* friars is likewife fpacious, and has good walls, but it is not fo magnificent as the former. The college of the Jefuits is fmall, but well built. To each of thefe three buildings are annexed fine large gardens, for the amufement, health, and ufe of the communities to which they belong. Some of the houfes in the town are built of ftone, but moft of them are of timber, though very neatly built. Each of the better fort of houfes has a door towards the ftreet, with a feat on each fide of it, for amufement and recreation in the morning and evening. The long ftreets are broad and ftrait, and divided at right angles by the fhort ones: fome are paved, but moft of them very uneven. The gates of the town are numerous; on the eaft fide of the town towards the river are five, two great and three leffer ones; and on the other fide are likewife feveral. The governor-general of *Canada*, when he is at *Montreal*, refides in the caftle, which the government hires for that purpofe of the family of *Vaudreuil*; but the governor of *Montreal* is obliged to buy or hire a houfe in town; though I was told, that the government contributed towards paying the rents.

In the town is a *Nunnery*, and without
its

its walls half a one; for though the laſt was quite ready, however, it had not yet been confirmed by the pope. In the firſt they do not receive every girl that offers herſelf; for their parents muſt pay about five hundred *ecus*, or crowns, for them. Some indeed are admitted for three hundred ecus, but they are obliged to ſerve thoſe who pay more than they. No poor girls are taken in.

THE king has erected a hoſpital for ſick ſoldiers here. The ſick perſon there is provided with every thing he wants, and the king pays twelve ſols every day for his ſtay, attendance, &c. The ſurgeons are paid by the king. When an officer is brought to this hoſpital, who is fallen ſick in the ſervice of the crown, he receives victuals and attendance gratis: but if he has got a ſickneſs in the execution of his private concerns, and comes to be cured here, he muſt pay it out of his own purſe. When there is room enough in the hoſpital, they likewiſe take in ſome of the ſick inhabitants of the town and country. They have the medicines, and the attendance of the ſurgeons, gratis, but muſt pay twelve ſols per day for meat, &c.

EVERY Friday is a market-day, when the country people come to the town with proviſions, and thoſe who want them muſt
ſupply

supply themselves on that day, because it is the only market-day in the whole week. On that day likewise a number of *Indians* come to town, to sell their goods, and buy others.

THE declination of the magnetic needle was here ten degrees and thirty-eight minutes, west. Mr. *Gillion*, one of the priests here, who had a particular taste for mathematicks and astronomy, had drawn a meridian in the garden of the seminary, which he said he had examined repeatedly by the sun and stars, and found to be very exact. I compared my compass with it, taking care, that no iron was near it, and found its declination just the same, as that which I have before mentioned.

ACCORDING to Monf. *Gillion*'s observations, the latitude of *Montreal* is forty-five degrees and twenty-seven minutes.

MONSR. *Pontarion*, another priest, had made thermometrical observations in *Montreal*, from the beginning of this year 1749. He made use of *Reaumur*'s thermometer, which he placed sometimes in a window half open, and sometimes in one quite open, and accordingly it will seldom mark the greatest degree of cold in the air. However, I shall give a short abstract of his observations for the winter months. In *January*

nuary the greatest cold was on the 18th day of the month, when the *Reaumurian* thermometer was twenty-three degrees below the freezing point. The least degree of cold was on the 31st of the same month, when it was just at the freezing point, but most of the days of this month it was from twelve to fifteen degrees below the freezing point. In *February* the greatest cold was on the 19th, and 25th, when the thermometer was fourteen degrees below the freezing point; and the least was on the 3d day of that month, when it rose eight degrees above the freezing point; but it was generally eleven degrees below it. In *March* the greatest cold was on the 3d, when it was ten degrees below the freezing point, and on the 22d, 23d, and 24th, it was mildest, being fifteen degrees above it: in general it was four degrees below it. In *April* the greatest degree of cold happened on the 7th, the thermometer being five degrees below the freezing point; the 25th was the mildest day, it being twenty degrees above the freezing point; but in general it was twelve degrees above it. These are the contents chiefly of Monf. *Pontarion*'s observations during those months; but I found, by the manner he made his observations, that the cold had every day been

from

from four to fix degrees greater, than he had marked it. He had likewife marked in his journal, that the ice in the river *St. Lawrence* broke on the 3d of *April* at *Montreal*, and only on the 20th day of that month at *Quebec*. On the 3d of *May* fome trees began to flower at *Montreal*, and on the 12th the hoary froft was fo great, that the trees were quite covered with it, as with fnow. The ice in the river clofe to this town is every winter above a *French* foot thick, and fometimes it is two of fuch feet, as I was informed by all whom I confulted on that head.

SEVERAL of the friars here told me, that the fummers were remarkably longer in *Canada*, fince its cultivation, than they ufed to be before; it begins earlier, and ends later. The winters on the other hand are much fhorter; but the friars were of opinion, that they were as hard as formerly, though they were not of the fame duration; and likewife, that the fummer at prefent was no hotter, than it ufed to be. The coldeft winds at *Montreal* are thofe from the north and north-weft.

Auguft the 2d. EARLY this morning we left *Montreal*, and went in a *bateau* on our journey to *Quebec*, in company with the fecond major of *Montreal*, M. *de Sermonville*.

ville. We fell down the river St. *Lawrence,* which was here pretty broad on our left; on the north-weſt ſide was the iſle of *Montreal,* and on the right a number of other iſles, and the ſhore. The iſle of *Montreal* was cloſely inhabited along the river; and it was very plain, and the riſing land near the ſhore conſiſted of pure mould, and was between three or four yards high. The woods were cut down along the riverſide, for the diſtance of an *Engliſh* mile. The dwelling-houſes were built of wood, or ſtone, indiſcriminately, and white-waſhed on the outſide. The other buildings, ſuch as barns, ſtables, *&c.* were all of wood. The ground next to the river was turned either into corn-fields, or meadows. Now and then we perceived churches on both ſides of the river, the ſteeples of which were generally on that ſide of the church, which looked towards the river, becauſe they are not obliged here to put the ſteeples on the weſt end of the churches. Within ſix *French* miles of *Montreal* we ſaw ſeveral iſlands of different ſizes on the river, and moſt of them were inhabited; and if ſome of them were without houſes on them, they were ſometimes turned into corn-fields, but generally into meadows. We ſaw no mountains, hills, rocks, or ſtones to-day, the country

country being flat throughout, and confifting of pure mould.

All the farms in *Canada* ftand feparate from each other, fo that each farmer has his poffeffions entirely diftinct from thofe of his neighbour. Each church, it is true, has a little village near it; but that confifts chiefly of the parfonage, a fchool for the boys and girls of the place, and of the houfes of tradefmen, but rarely of farm-houfes; and if that was the cafe, yet their fields were feparated. The farm-houfes hereabouts are generally built all along the rifing banks of the river, either clofe to the water or at fome diftance from it, and about three or four *arpens* from each other. To fome farms are annexed fmall orchards; but they are in general without them; however, almoft every farmer has a kitchen-garden.

I have been told by all thofe who have made journies to the fouthern parts of *Canada*, and to the river *Miffifippi*, that the woods there abound with peach-trees, which bear excellent fruit, and that the *Indians* of thofe parts fay, that thofe trees have been there fince times immemorial.

The farm-houfes are generally built of ftone, but fometimes of timber, and have three or four rooms. The windows are

feldom

seldom of glass, but most frequently of paper. They have iron stoves in one of the rooms, and chimnies in the rest. The roofs are covered with boards. The crevices and chinks are filled up with clay. The other buildings are covered with straw.

There are several *Crosses* put up with the road side, which is parallel to the shores of the river. These crosses are very common in *Canada*, and are put up to excite devotion in the travellers. They are made of wood, five or six yards high, and proportionally broad. In that side which looks towards the road is a square hole, in which they place an image of our Saviour, the cross, or of the holy Virgin, with the child in her arms; and before that they put a piece of glass, to prevent its being spoiled by the weather. Those crosses which are not far from churches, are very much adorned, and they put up about them all the instruments which they think the *Jews* employed in crucifying our Saviour, such as a hammer, tongs, nails, a flask of vinegar, and perhaps many more than were really made use of. A figure of the cock, which crowed when *St. Peter* denied our Lord, is commonly put at the top of the cross.

The country on both sides was very delightful

lightful to day, and the fine state of its cultivation, added greatly to the beauty of the scene. It could really be called a village, beginning at *Montreal*, and ending at *Quebec*, which is a distance of more than one hundred and eighty miles; for the farm-houses are never above five arpens, and sometimes but three, asunder, a few places excepted. The prospect is exceedingly beautiful, when the river goes on for some miles together in a strait line, because it then shortens the distances between the houses, and makes them form exactly one continued village.

All the women in the country, without exception, wear caps of some kind or other. Their jackets are short, and so are their petticoats, which scarce reach down to the middle of their legs; and they have a silver cross hanging down on the breast. In general they are very laborious; however, I saw some, who, like the *English* women in the colonies, did nothing but prattle all the day. When they have any thing to do within doors, they (especially the girls) commonly sing songs, in which the words *Amour* and *Cœur* are very frequent. In the country it is usual, that when the husband receives a visit from persons of rank, and dines with them, his wife stands behind and serves him; but in the towns, the ladies are more diftinguifhed, and would willingly affume an equal, if not a fuperior, power to their hufbands. When they go out of doors they wear long cloaks, which cover all their other clothes, and are either grey, brown, or blue. The men fometimes make ufe of them, when they are obliged to go into the rain. The women have the advantage of being in a *defhabille* under thefe cloaks, without any body's perceiving it.

WE fometimes faw wind-mills near the farms. They were generally built of ftone, with a roof of boards, which, together with its flyers, could be turned to the wind occafionally.

THE breadth of the river was not always equal to-day; in the narroweft place, it was about a quarter of an *Englifh* mile broad; in other parts it was near two *Englifh* miles. The fhore was fometimes high and fteep, and fometimes low, or floping.

AT three o'clock this afternoon we paffed by the river, which falls into the river St. *Lawrence*, and comes from lake *Champlain*, in the middle of which latter is a large ifland. The yachts which go between *Montreal* and *Quebec*, go on the fouth-eaft fide of this ifland, becaufe it is

deeper

deeper there; but the boats prefer the north-west side, because it is nearer, and yet deep enough for them. Besides this island there are several more hereabouts, which are all inhabited. Somewhat further, the country on both sides the river is uninhabited, till we come to the *Lac St. Pierre*; because it is so low, as to be quite overflowed at certain times of the year. To make up for this deficiency, the country, I am told, is as thickly inhabited further from the river, as we found it along the banks of the river.

Lac St. Pierre is a part of the river St. *Lawrence*, which is so broad that we could hardly see any thing but sky and water before us, and I was every where told, that it is seven *French* miles long, and three broad. From the middle of this lake as it is called, you see a large high country in the west, which appears above the woods. In the lake are many places covered with a kind of rush, or *Scirpus palustris*, *Linn.* There are no houses in sight on either side of the lake, because the land is rather too low there; and in spring the water rises so high, that they may go with boats between the trees. However, at some distance from the shores, where the ground is higher, the farms are close together. We saw no islands in the lake

lake this afternoon, but the next day we met with some.

Late in the evening we left lake St. *Pierre*, and rowed up a little river called *Riviere de Loup*, in order to come to a house where we might pass the night. Having rowed about an *English* mile, we found the country inhabited on both sides of the river. Its shores are high; but the country in general is flat. We passed the night in a farm-house. The territory of *Montreal* extends to this place; but here begins the jurisdiction of the governor of *Trois Rivieres*, to which place they reckon eight *French* miles from hence.

August the 3d. At five o'clock in the morning we set out again, and first rowed down the little river till we came into the lake St. *Pierre*, which we went downwards. After we had gone a good way, we perceived a high chain of mountains in the north-west, which were very much elevated above the low, flat country. The north-west shore of lake St. *Pierre* was now in general very closely inhabited; but on the south-east side we saw no houses, and only a country covered with woods, which is sometimes said to be under water, but behind which there are, as I am told, a great number of farms. Towards the end

end of the lake, the river went into its proper bounds again, being not above a mile and a half broad, and afterwards it grows still narrower. From the end of Lake St. *Pierre* to *Trois Rivieres*, they reckon three *French* miles, and about eleven o'clock in the morning we arrived at the latter place, where we attended divine service.

Trois Rivieres, is a little market town, which had the appearance of a large village; it is however reckoned among the three great towns of *Canada*, which are *Quebec*, *Montreal*, and *Trois Riveres*. It is said to ly in the middle between the two first, and thirty *French* miles distant from each. The town is built on the north side of the river St. *Lawrence*, on a flat, elevated sand, and its situation is very pleasant. On one side the river passes by, which is here an *English* mile and a half broad. On the other side, are fine corn-fields, though the soil is very much mixed with sand. In the town are two churches of stone, a nunnery, and a house for the friars of the order of St. *Francis*. This town is likewise the seat of the third governor in *Canada*, whose house is likewise of stone. Most of the other houses are of timber a single story high, tolerably well built, and stand very much asunder; and the streets are crooked. The shore here consists

consists of sand, and the rising grounds along it are pretty high. When the wind is very violent here, it raises the sand, and blows it about the streets, making it very troublesome to walk in them. The nuns, which are about twenty-two in number, are reckoned very ingenious in all kinds of needle-work. This town formerly flourished more than any other in *Canada*, for the *Indians* brought their goods to it from all sides; but since that time they go to *Montreal* and *Quebec*, and to the *English*, on account of their wars with the *Iroquese*, or Five Nations, and for several other reasons, so that this town is at present very much reduced by it. Its present inhabitants live chiefly by agriculture, though the neighbouring iron-works may serve in some measure to support them. About an *English* mile below the town, a great river falls into the river St. *Lawrence*, but first divides into three branches, so that it appears as if three rivers disembogued themselves there. This has given occasion to call the river and this town, *Trois Rivieres (the Three Rivers)*.

THE tide goes about a *French* mile above *Trois Rivieres*, though it is so trifling as to be hardly observable. But about the equinoxes, and at the new moons and full moons in spring and autumn, the difference between the highest

highest and lowest water is two feet. Accordingly the tide in this river goes very far up, for from the above mentioned place to the sea they reckon about a hundred and fifty *French* miles.

WHILST my company were resting, I went on horseback to view the iron-work. The country which I passed through was pretty high, sandy, and generally flat. I saw neither stones nor mountains here.

THE *iron-work*, which is the only one in this country, lies three miles to the west of *Trois Rivieres*. Here are two great forges, besides two lesser ones to each of the great ones, and under the same roof with them. The bellows were made of wood, and every thing else, as it is in *Swedish* forges. The melting ovens stand close to the forges, and are the same as ours. The ore is got two *French* miles and a half from the iron works, and is carried thither on sledges. It is a kind of moor ore *, which lies in veins, within six inches or a foot from the surface of the ground. Each vein is from six to eighteen inches deep, and below it is a white sand.

The

---

* *Tophus Tubalcaini, Linn. Syst. Nat.* III. p. 187, n. 5. *Minera ferri subaquosa nigro cærulescens. Wall. Mineral.* p. 263. *Germ.* Ed. p. 340. n. 3. *Iron ockres* in the shape of crusts, are sometimes cavernous, as the *Brush ore*. *Forster's Mineral.* p. 48.

The veins are furrounded with this fand on both fides, and covered at the top with a thin mould. The ore is pretty rich and lies in loofe lumps in the veins, of the fize of two fifts, though there are a few which are near eighteen inches thick. Thefe lumps are full of holes, which are filled with ockre. The ore is fo foft that it may be crufhed betwixt the fingers. They make ufe of a grey lime-ftone, which is broke in the neighbourhood, for promoting the fufibility of the ore; to that purpofe they likewife employ a clay marle, which is found near this place. Charcoals are to be had in great abundance here, becaufe all the country round this place is covered with woods, which have never been ftirred. The charcoals from ever-green trees, that is, from the fir kind, are beft for the forge, but thofe of deciduous trees are beft for the fmelting oven. The iron which is here made, was to me defcribed as foft, pliable, and tough, and is faid to have the quality of not being attacked by ruft fo eafily as other iron; and in this point there appears a great difference between the *Spanifh* iron and this in fhip-building. This iron-work was firft founded in 1737, by private perfons, who afterwards ceded it to the king; they caft cannon and mortars here, of different fizes, sizes, iron stoves which are in use all over *Canada*, kettles, &c. not to mention the bars which are made here. They have likewise tried to make steel here, but cannot bring it to any great perfection, because they are unacquainted with the best manner of preparing it. Here are many officers and overseers, who have very good houses, built on purpose for them. It is agreed on all hands, that the revenues of the iron-work do not pay the expences which the king must every year be at in maintaining it. They lay the fault on the bad state of population, and say that the few inhabitants in the country have enough to do with agriculture, and that it therefore costs great trouble and large sums, to get a sufficient number of workmen. But however plausible this may appear, yet it is surprizing that the king should be a loser in carrying on this work; for the ore is easily broken, very near the iron-work, and very fusible. The iron is good, and can be very conveniently dispersed over the country. This is moreover the only iron-work in the country, from which every body must supply himself with iron tools, and what other iron he wants. But the officers and servants belonging to the iron-work, appear to be in very affluent circumstances. A river runs down from the iron-work, into the river St. *Lawrence*, by which all the iron can be sent in boats throughout the country at a low rate. In the evening I returned again to *Trois Rivieres*.

August the 4th. AT the dawn of day we left this place and went on towards *Quebec*. We found the land on the north side of the river somewhat elevated, sandy, and closely inhabited along the water side. The south-east shore, we were told, is equally well inhabited; but the woods along that shore prevented our seeing the houses, which are built further up in the country, the land close to the river being so low as to be subject to annual inundations. Near *Trois Rivieres*, the river grows somewhat narrow; but it enlarges again, as soon as you come a little below that place, and has the breadth of above two *English* miles.

As we went on, we saw several churches of stone, and often very well built ones. The shores of the river are closely inhabited for about three quarters of an *English* mile up the country; but beyond that the woods and the wilderness encrease. All the rivulets falling into the river St. *Lawrence* are likewise well inhabited on both sides. I observed throughout *Canada*, that the cultivated

vated lands ly only along the river St. *Lawrence*, and the other rivers in the country, the environs of towns excepted, round which the country is all cultivated and inhabited within the diſtance of twelve or eighteen *Engliſh* miles. The great iſlands in the river are likewiſe inhabited.

THE ſhores of the river now became higher, more oblique and ſteep, however they conſiſted chiefly of earth. Now and then ſome rivers or great brooks fall into the river St. *Lawrence*, among which one of the moſt conſiderable is the *Riviere Puante*, which unites on the ſouth-eaſt ſide with the St. *Lawrence*, about two *French* miles below *Trois Rivieres*, and has on its banks, a little way from its mouth, a town called *Becancourt* which is wholly inhabited by *Abenakee Indians*, who have been converted to the *Roman catholic* religion, and have *Jeſuits* among them. At a great diſtance, on the north-weſt ſide of the river, we ſaw a chain of very high mountains, running from north to ſouth, elevated above the reſt of the country, which is quite flat here without any remarkable hills.

HERE were ſeveral lime-kilns along the river; and the lime-ſtone employed in them is broke in the neighbouring high grounds. It is compact and grey, and the lime it yields is pretty white.

THE fields here are generally sown with wheat, oats, maize, and peafe. Gourds and water-melons are planted in abundance near the farms.

A Humming bird *(Trochilus Colubris)* flew among the bufhes, in a place where we landed to day. The *French* call it *Oifeau mouche*, and fay it is pretty common in *Canada*; and I have feen it fince feveral times at *Quebec*.

ABOUT five o'clock in the afternoon we were obliged to take our night's lodgings on fhore, the wind blowing very ftrong againft us, and being attended with rain. I found that the nearer we came to *Quebec*, the more open and free from woods was the country. The place where we paffed the night, is diftant from *Quebec* twelve *French* miles.

THEY have a very peculiar method of catching fifh near the fhore here. They place hedges along the fhore, made of twifted oziers, fo clofe that no fifh can get through them, and from one foot to a yard high, according to the different depth of the water. For this purpofe they choofe fuch places where the water runs off during the ebb, and leaves the hedges quite dry. Within this inclofure they place feveral weels, or fifh-traps, in the form of cylinders, but broader below. They are placed upright, and

and are about a yard high, and two feet and a half wide: on one side near the bottom is an entrance for the fishes, made of twigs, and sometimes of yarn made into a net. Opposite to this entrance, on the other side of the weel, looking towards the lower part of the river, is another entrance, like the first, and leading to a box of boards about four foot long, two deep, and two broad. Near each of the weels is a hedge, leading obliquely to the long hedge, and making an acute angle with it. This latter hedge is made in order to lead the fish into the trap, and it is placed on that end of the long hedge which looks towards the upper part of the river; now when the tide comes up the river, the fish, and chiefly the eels, go up with it along the river side; when the water begins to ebb, the fish likewise go down the river, and meeting with the hedges, they swim along them, till they come through the weels into the boxes of boards, at the top of which there is a hole with a cover, through which the fish could be taken out. This apparatus is chiefly made on account of the eels. In some places hereabouts they place nets instead of the hedges of twigs.

The shores of the river now consisted no more of pure earth; but of a species of slate. They are very steep and nearly perpendicular

pendicular here, and the flates of which they confift are black, with a brown caft; and divifible into thin fhivers, no thicker than the back of a knife. Thefe flates moulder as foon as they are expofed to the open air, and the fhore is covered with grains of fmall fand, which are nothing but particles of fuch mouldered flates. Some of the ftrata run horizontal, others obliquely, dipping to the fouth and rifing to the north, and fometimes the contrary way. Sometimes they form bendings like large femicircles: fometimes a perpendicular line cuts off the ftrata, to the depth of two feet; and the flates on both fides of the line from a perpendicular and fmooth wall. In fome places hereabouts, they find amongft the flates, a ftratum about four inches thick of a grey, compact, but pretty foft limeftone, of which the *Indians* for many centuries have made, and the *French* at prefent ftill make, tobacco-pipes*.

Auguft the 5th. THIS morning, we continued our journey by rowing, the contrary wind hindering us from failing. The appearance

---

* This lime-ftone, feems to be a marle, or rather a kind of ftone-marle: for there is a whitifh kind of it in the *Krim-Tartary*, and near *Stiva* or *Thebes*, in *Greece*, which is employed by the *Turks* and *Tartars* for making heads of pipes, and that from the firft place is called *Keffekil*, and in the latter, *Sea-Scum*: it may be very eafily cut, but grows harder in time. F.

pearance of the shores, was the same as yesterday; they were high, pretty steep, and quite perpendicular; and consisted of the black slate before described. The country at the top was a plain without eminences, and closely inhabited along the river, for about the space of an *English* mile and a half in-land. Here are no islands in this part of the river, but several stony places, perceptible at low water only, which have several times proved fatal to travellers. The breadth of the river varies; in some parts it was a little more than three quarters of a mile, in others half a mile, and in some above two miles. The inhabitants made use of the same method of catching eels along the shores here, as that which I have just before mentioned. In many places they make use of nets made of osiers instead of the hedge.

Bugs (*Cimex lectularius*) abound in *Canada*; and I met with them in every place where I lodged, both in the towns and country, and the people know of no other remedy for them than patience.

The Crickets (*Gryllus domesticus*) are also abundant in *Canada*, especially in the country, where these disagreeable guests lodge in the chimnies; nor are they uncommon in the towns. They stay here both summer and

and winter, and frequently cut clothes in pieces for paftime.

The Cockroaches (*Blatta orientalis*) have never been found in the houfes here.

The fhores of the river grow more floping as you come nearer to *Quebec*. To the northward appears a high ridge of mountains. About two *French* miles and a half from *Quebec*, the river becomes very narrow, the fhores being within the reach of a mufket fhot from each other. The country on both fides was floping, hilly, covered with trees, and had many fmall rocks; the fhore was ftony. About four o'clock in the afternoon we happily arrived at *Quebec*. The city does not appear till one is clofe to it, the profpect being intercepted by a high mountain on the fouth fide. However, a part of the fortifications appears at a good diftance, being fituate on the fame mountain. As foon as the foldiers, who were with us, faw *Quebec*, they called out, that all thofe who had never been there before, fhould be ducked, if they did not pay fomething to releafe themfelves. This cuftom even the governor-general of *Canada* is obliged to fubmit to, on his firft journey to *Montreal*. We did not care when we came in fight of this town to be exempted from this old cuftom, which is very advantageous

vantageous to the rowers, as it enables them to spend a merry evening on their arrival at *Quebec*, after their troublesome labour.

IMMEDIATELY after my arrival, the officer who had accompanied me from *Montreal*, led me to the palace of the then vice-govenor-general of *Canada*, the marquis *la Galissonniere*, a nobleman of uncommon qualities, who behaved towards me with extraordinary goodness, during the time he staid in this country. He had already ordered some apartments to be got ready for me, and took care to provide me with every thing I wanted; besides honouring me so far to invite me to his table, almost every day I was in town.

August the 6th. *Quebec*, the chief city in *Canada*, lies on the western shore of the river St. *Lawrence*, close to the water's edge, on a neck of land, bounded by that river on the east side, and by the river St. *Charles* on the north side; the mountain, on which the town is built, rises still higher on the south side, and behind it begin great pastures; and the same mountain likewise extends a good way westward. The city is distinguished into the lower and the upper[*]. The lower lies on the river, east-

[*] *La haute Ville & la basse Ville.*

ward of the upper. The neck of land,
I mentioned before, was formed by the
dirt and filth, which had from time to time
been accumulated there, and by a rock
which lay that way, not by any gradual
diminution of the water. The upper city
lies above the other, on a high hill, and
takes up five or fix times the fpace of the
lower, though it is not quite fo populous.
The mountain, on which the upper city is
fituated, reaches above the houfes of the
lower city. Notwithftanding the latter are
three or four ftories high, and the view,
from the palace, of the lower city (part of
which is immediately under it) is enough
to caufe a fwimming of the head. There
is only one eafy way of getting to the upper city, and there part of the mountain
has been blown up. This road is very
fteep, notwithftanding it is made winding
and ferpentine. However, they go up and
down it in carriages, and with waggons.
All the other roads up the mountain are
fo fteep, that it is very difficult to climb to
the top by them. Moft of the merchants
live in the lower city, where the houfes
are built very clofe together. The ftreets
in it are narrow, very rugged, and almoft
always wet. There is likewife a church,
and a fmall market-place. The upper city

is

is inhabited by people of quality, by several perſons belonging to the different offices, by tradeſmen, and others. In this part are the chief buildings of the town, among which the following are worthy particular notice.

I. The *Palace* is ſituated on the weſt or ſteepeſt ſide of the mountain, juſt above the lower city. It is not properly a palace, but a large building of ſtone, two ſtories high, extending north and ſouth. On the weſt ſide of it is a court-yard, ſurrounded partly with a wall, and partly with houſes. On the eaſt ſide, or towards the river, is a gallery as long as the whole building, and about two fathom broad, paved with ſmooth flags, and included on the outſides by iron rails, from whence the city and the river exhibit a charming proſpect. This gallery ſerves as a very agreeable walk after dinner, and thoſe who come to ſpeak with the governor-general wait here till he is at leiſure. The palace is the lodging of the governor-general of *Canada*, and a number of ſoldiers mount the guard before it, both at the gate and in the court-yard; and when the governor, or the biſhop, comes in or goes out, they muſt all appear in arms, and beat the drum. The governor-general has his own chapel where he hears prayers; however, he often goes to mass at the church of the *Recolets* *, which is very near the palace.

II. The *Churches* in this town are seven or eight in number, and all built of stone.

1. The *Cathedral* church is on the right hand, coming from the lower to the upper city, somewhat beyond the bishop's house. The people were at present employed in ornamenting it. On its west side is a round steeple, with two divisions, in the lower of which are some bells. The pulpit, and some other parts within the church, are gilt. The seats are very fine.

2. The *Jesuits* church is built in the form of a cross, and has a round steeple. This is the only church that has a clock, and I shall mention it more particularly below.

3. The *Recolets* church is opposite the gate of the palace, on the west side, looks well, and has a pretty high pointed steeple, with a division below for the bells.

4. The church of the *Ursulines* has a round spire.

5. The church of the hospital.

6. The bishop's chapel.

7. The

---

* A kind of *Franciscan* friars, called *Ordo Sti. Francisci strictioris observantiæ*.

7. THE church in the lower city was built in 1690, after the town had been delivered from the *Englifh*, and is called *Notre Dame de la Victoire*. It has a fmall fteeple in the middle of the roof, fquare at the bottom, and round at the top.

8. THE little chapel of the governor-general, may likewife be ranked amongft thefe churches.

III. THE bifhop's houfe is the firft, on the right hand, coming from the lower to the upper town. It is a fine large building, furrounded by an extenfive courtyard and kitchen-garden on one fide, and by a wall on the other.

IV. THE college of the Jefuits, which I will defcribe more particularly. It has a much more noble appearance, in regard to its fize and architecture, than the palace itfelf, and would be proper for a palace if it had a more advantageous fituation. It is about four times as large as the palace, and is the fineft building in town. It ftands on the north fide of a market, on the fouth fide of which is the cathedral.

V. THE houfe of the Recolets lies to the weft, near the palace and directly over againft it, and confifts of a fpacious building, with a large orchard, and kitchen-garden. The houfe is two ftories high.

In each story is a narrow gallery with rooms and halls on one, or both sides.

VI. The *Hôtel de Dieu*, where the sick are taken care of, shall be described in the sequel. The nuns, that serve the sick, are of the *Augustine* order.

VII. The house of the clergy * is a large building, on the north-east side of the cathedral. Here is on one side a spacious court, and on the other, towards the river, a great orchard, and kitchen-garden. Of all the buildings in the town none has so fine a prospect as that in the garden belonging to this house, which lies on the high shore, and looks a good way down the river. The Jesuits on the other hand have the worst, and hardly any prospect at all from their college; nor have the Recolets any fine views from their house. In this building all the clergy of *Quebec* lodge with their superior. They have large pieces of land in several parts of *Canada*, presented to them by the government, from which they derive a very plentiful income.

VIII. The convent of the *Ursuline* nuns shall be mentioned in the sequel.

These are all the chief public buildings in the town, but to the north-west, just before the town, is

* *Le Seminaire.*

IX.

IX. The house of the intendant, a public building, whose size makes it fit for a palace. It is covered with tin, and stands in a second lower town, situated southward upon the river St. *Charles*. It has a large and fine garden on its north side. In this house all the deliberations concerning this province, are held; and the gentlemen who have the management of the police and the civil power meet here, and the intendant generally presides. In affairs of great consequence the governor-general is likewise here. On one side of this house is the store-house of the crown, and on the other the prison.

Most of the houses in *Quebec* are built of stone, and in the upper city they are generally but one story high, the public buildings excepted. I saw a few wooden houses in the town, but they must not be rebuilt when decayed. The houses and churches in the city are not built of bricks, but the black lime-slates of which the mountain consists, whereon *Quebec* stands. When these lime-slates are broke at a good depth in the mountain, they look very compact at first, and appear to have no shivers, or *lamellæ*, at all; but after being exposed a while to the air, they separate into thin leaves. These slates are soft, and easily cut;

cut; and the city-walls, together with the garden-walls, confift chiefly of them. The roofs of the public buildings are covered with common flates, which are brought from *France*, becaufe there are none in *Canada*.

THE flated roofs have for fome years withftood the changes of air and weather, without fuffering any damage. The private houfes have roofs of boards, which are laid parallel to the fpars, and fometimes to the eaves, or fometimes obliquely. The corners of houfes are made of a grey fmall grained lime-ftone, which has a ftrong fmell, like the *ftinkftone**, and the windows are generally enchafed with it. This lime-ftone is more ufeful in thofe places than the lime-flates, which always fhiver in the air. The outfides of the houfes are generally whitewafhed. The windows are placed on the inner fide of the walls; for they have fometimes double windows in winter. The middle roof has two, or at moft three fpars, covered with boards only. The rooms are warmed in winter by fmall iron ftoves, which are removed in fummer. The floors are very dirty in every houfe, and have all

the

---

* *Nitrum fuillum. Linn.* Syft. III. p. 86. Lapis fuillus prifmaticus Waller. Mineral. p. 59. a. 1. *Stink-ftone*, Forfter's Introd. to Mineralogy. p. 40.

the appearance of being cleaned but once every year.

The *Powder magazine* stands on the summit of the mountain, on which the city is built, and southward of the palace.

The streets in the upper city have a sufficient breadth, but are very rugged, on account of the rock on which it lies; and this renders them very disagreeable and troublesome, both to foot-passengers and carriages. The black lime-slates basset out and project every where into sharp angles, which cut the shoes in pieces. The streets cross each other at all angles, and are very crooked.

The many great orchards and kitchen-gardens, near the house of the Jesuits, and other public and private buildings, make the town appear very large, though the number of houses it contains is not very considerable. Its extent from south to north is said to be about six hundred toises, and from the shore of the river along the lower town, to the western wall between three hundred and fifty, and four hundred toises. It must be here observed, that this space is not yet wholly inhabited; for on the west and south side, along the town walls, are large pieces of land without any buildings on them, and destined to be built upon in future times, when the number of inhabitants will be encreafed in *Quebec*.

THE bifhop, whofe fee is in the city, is the only bifhop in *Canada*. His diocefe extends to *Louifiana*, on the *Mexican* gulf fouthward, and to the fouth-feas weftward.

No bifhop, the pope excepted, ever had a more extenfive diocefe. But his fpiritual flock is very inconfiderable at fome diftance from *Quebec*, and his fheep are often many hundred miles diftant from each other.

Quebec is the only fea-port and trading town in all *Canada*, and from thence all the produce of the country is exported. The port is below the town in the river, which is there about a quarter of a *French* mile broad, twenty-five fathoms deep, and its ground is very good for anchoring. The fhips are fecured from all ftorms in this port; however, the north-eaft wind is the worft, becaufe it can act more powerfully. When I arrived here, I reckoned thirteen great and fmall veffels, and they expected more to come in. But it is to be remarked, that no other fhips than *French* ones can come into the port, though they may come from any place in *France*, and likewife from the *French* poffeffions in the

Weft-

West-Indies. All the foreign goods, which are found in *Montreal*, and other parts of *Canada*, muſt be taken from hence. The *French* merchants from *Montreal* on their ſide, after making a ſix months ſtay among ſeveral *Indian* nations, in order to purchaſe ſkins of beaſts and furrs, return about the end of *Auguſt*, and go down to *Quebec* in *September* or *October*, in order to ſell their goods there. The privilege of ſelling the imported goods, it is ſaid, has vaſtly enriched the merchants of *Quebec*; but this is contradicted by others, who allow that there are a few in affluent circumſtances, but that the generality poſſeſs no more than is abſolutely neceſſary for their bare ſubſiſtence, and that ſeveral are very much in debt, which they ſay is owing to their luxury and vanity. The merchants dreſs very finely, and are extravagant in their repaſts; and their ladies are every day in full dreſs, and as much adorned as if they were to go to court.

The town is ſurrounded on almoſt all ſides by a high wall, and eſpecially towards the land. It was not quite completed when I was there, and they were very buſy in finiſhing it. It is built of the above mentioned black lime-ſlate, and of a dark-grey ſandſtone. For the corners of the gates they have

have employed a grey lime-stone. They have not made any walls towards the water side, but nature seems to have worked for them, by placing a rock there which it is impossible to ascend. All the rising land thereabouts is likewise so well planted with cannon, that it seems impossible for an enemy's ships or boats to come to the town without running into imminent danger of being sunk. On the land side the town is likewise guarded by high mountains so that nature and art have combined to fortify it.

Quebec was founded by its former governor, *Samuel de Champlain*, in the year 1608. We are informed by history, that its rise was very slow. In 1629 towards the end of *July* it was taken by two *Englishmen Lewis* and *Thomas Kerk*, by capitulation, and surrendered to them by the above mentioned *de Champlain*. At that time, *Canada* and *Quebec* were wholly destitute of provisions, so that they looked upon the *English* more as their deliverers, than their enemies. The abovementioned *Kerks*, were the brothers of the *English* admiral *David Kerk*, who lay with his fleet somewhat lower in the river. In the year 1632, the *French* got the town of *Quebec*, and all *Canada* returned to them by

by the peace. It is remarkable, that the *French* were doubtful whether they should reclaim *Canada* from the *English* or leave it to them. The greater part were of opinion that to keep it would be of no advantage to *France*, because the country was cold; and the expences far exceeded its produce; and because *France* could not people so extensive a country without weakening herself, as *Spain* had done before. That it was better to keep the people in *France*, and employ them in all sorts of manufactures, which would oblige the other *European* powers who have colonies in *America* to bring their raw goods to *French* ports, and take *French* manufactures in return. Those on the other hand who had more extensive views knew that the climate was not so rough as it had been represented. They likewise believed that that which caused the expences was a fault of the company, because they did not manage the country well. They would not have many people sent over at once, but little by little, so that *France* might not feel it. They hoped that this colony would in future times make *France* powerful, for its inhabitants would become more and more acquainted with the herring, whale, and cod fisheries, and likewise with
the

the taking of seals; and that by this means *Canada* would become a school for training up seamen. They further mentioned the several sorts of furrs, the conversion of the *Indians*, the ship-building, and the various uses of the extensive woods. And lastly that it would be a considerable advantage to *France*, even though they should reap no other benefit, to hinder by this means the progress of the *English* in *America*, and of their encreasing power, which would otherwise become insupportable to *France*; not to mention several other reasons. Time has shewn that these reasons were the result of mature judgment, and that they laid the foundation to the rise of *France*. It were to be wished that we had been of the same opinion in *Sweden*, at a time when we were actually in possession of *New Sweden*, the finest and best province in all *North America*, or when we were yet in a condition to get the possession of it. Wisdom and foresight does not only look upon the present times, but even extends its views to futurity.

In the year 1663 at the beginning of *February*, the great earthquake was felt in *Quebec* and a great part of *Canada*, and there are still some vestiges of its effects at that time; however, no lives were lost.

On

On the 16th of *October* 1690, *Quebec* was besieged by the *English* general *William Phips*, who was obliged to retire a few days after with great loss. The *English* have tried several times to repair their losses, but the river St. *Lawrence* has always been a very good defence for this country. An enemy, and one that is not acquainted with this river, cannot go upwards in it, without being ruined; for in the neighbourhood of *Quebec*, it abounds with hidden rocks, and has strong currents in some places, which oblige the ships to make many windings.

The name of *Quebec* it is said is derived from a *Norman* word, on account of its situation on a neck or point of land. For when one comes up in the river by *l'Isle d'Orleans*, that part of the river St. *Lawrence* does not come in sight, which lies above the town, and it appears as if the river St. *Charles* which lies just before, was a continuation of the St. *Lawrence*. But on advancing further the true course of the river comes within sight, and has at first a great similarity to the mouth of a river or a great bay. This has given occasion to a sailor, who saw it unexpectedly, to cry out in his provincial dialect *Que bec* [*], that is, *what a point of land!* and from hence it is thought

---
[*] Meaning *Quel bec*.

thought the city obtained its name. Others derive it from the *Algonkin* word *Quebego* or *Quebec* fignifying that which grows narrow, becaufe the river becomes narrower as it comes nearer to the town.

THE river St. *Lawrence*, is exactly a quarter of a *French* mile, or three quarters of an *English* mile broad at *Quebec*. The falt water never comes up to the town in it, and therefore the inhabitants can make ufe of the water in the river for their kitchens, &c. All accounts agree that notwithftanding the breadth of this river, and the violence of its courfe, efpecially during ebb, it is covered with ice during the whole winter, which is ftrong enough for walking, and a carriage may go over it. It is faid to happen frequently that, when the river has been open in *May*, there are fuch cold nights in this month, that it freezes again, and will bear walking over. This is a clear proof of the intenfenefs of the froft here, efpecially when one confiders that which I fhall mention immediately after, about the ebbing and flowing of the tide in this river. The greateft breadth of the river at its mouth, is computed to be twenty-fix *French* miles or feventy-eight *English* miles, though the boundary between the fea, and the river cannot well be afcertained as the latter gradually loofes itfelf in, and unites with the

the former. The greatest part of the water contained in the numerous lakes of *Canada*, four or five of which are like large seas, is forced to disembogue into the sea by means of this river alone. The navigation up this river from the sea is rendered very dangerous by the strength of the current, and by the number of sand-banks, which often arise in places where they never were before. The *English* have experienced this formation of new sands once or twice, when they intend to conquer *Canada*. Hence the *French* have good reasons to look upon the river as a barrier to *Canada**.

THE tide goes far beyond *Quebec* in the river St. *Lawrence*, as I have mentioned above. The difference between high and low water is generally between fifteen and sixteen feet, *French* measure; but with the new and full moon, and when the wind is likewise favourable, the difference is seventeen or eighteen feet, which is indeed very considerable.

* The river St. *Lawrence*, was no more a barrier to the victorious *British* fleets in the last war, nor were the fortifications of *Quebec* capable to withstand the gallant attacks of their land army, which disappointed the good *Frenchmen* in *Canada* of their too sanguine expectations, and at present, they are rather happy at this change of fortune, which has made them subjects of the *British* sceptre, whose mild influence they at present enjoy. F.

VOL. III.　　H　　*August*

August the 7th. *Ginseng* is the current *French* name in *Canada*, of a plant, the root of which, has a very great value in *China**. It has been growing since times immemorial in the *Chinese Tartary* and in *Corea*, where it is annually collected and brought to *China*. Father *Du Halde* says, it is the most precious, and the most useful of all the plants in eastern *Tartary*, and attracts, every year, a number of people into the deserts of that country. The *Mantechoux-Tartars* call it *Orhota*, that is the most noble, or the queen of plants†. The *Tartars* and *Chinese* praise it very much, and ascribe to it the power of curing several dangerous diseases, and that of restoring to the body new strength, and supplying the loss caused by the exertion of the mental, and corporeal faculties. An ounce of *Ginseng* bears the surprizing price of seven or eight ounces of silver at *Peking*. When the *French* botanists in *Canada* first saw a figure of it, they remembered to have seen

---

* Botanists know this plant by the name of *Panax quinquefolium*, foliis ternatis quinatis Linn. Mat. Med. § 116. Sp. plant. p. 15. 12. *Gronov.* Fl. *Virg.* p. 147. See likewise *Catesby's* Nat. Hist of *Carolina*. Vol. III. p. 16. t. 16. *Lafitau* Ginf. 51. t. 1. Father *Charlevoix* Hist. de la Nouvelle France. Tom. IV. p. 308. fig. XIII. and Tom. V. p. 24.

† *Peter Osbeck's* voyage to *China*, Vol. I. p. 223.

a fimiliar plant in this country. They were confirmed in their conjecture by confidering that feveral fettlements in *Canada*, ly under the fame latitude with thofe parts of the *Chinefe Tartary*, and *China*, where the true *Ginfeng* grows wild. They fucceeded in their attempt, and found the fame *Ginfeng* wild and abundant in feveral parts of *North-America*, both in *French* and *Englifh* plantations, in plain parts of the woods. It is fond of fhade, and of a deep rich mould, and of land which is neither wet nor high. It is not every where very common, for fometimes one may fearch the woods for the fpace of feveral miles without finding a fingle plant of it; but in thofe fpots where it grows it is always found in great abundance. It flowers in *May* and *June*, and its berries are ripe at the end of *Auguft*. It bears tranfplanting very well, and will foon thrive in its new ground. Some people here, who have gathered the berries, and put them into their kitchen gardens, told me that they lay one or two years in the ground without coming up. The *Iroquefe*, or Five (Six) Nations, call the *Ginfeng* roots *Garangtoging*, which it is faid fignifies *a child*, the roots bearing a faint refemblance to it: but others are of opinion that they mean the thigh and leg by it, and the roots look pretty like it. The *French* ufe this root for curing the afthma, as a ftomachic, and to promote fertility in woman. The trade which is carried on with it here is very brifk; for they gather great quantities of it, and fend them to *France*, from whence they are brought to *China*, and fold there to great advantage *. It is faid the merchants in *France* met with amazing fuccefs in this trade at the firft outfet, but by continuing to fend the *Ginfeng* over to *China*, its price is fallen confiderably there, and confequently in *France* and *Canada*; however, they ftill find their account in it. In the fummer of 1748, a pound of *Ginfeng* was fold for fix Francs, or Livres, at *Quebec*; but its common price here is one hundred Sols, or five Livres. During my ftay in *Canada*, all the merchants at *Quebec* and *Montreal*, received orders from their correfpondents in *France* to fend over a quantity of *Ginfeng*, there being an uncommon demand for it this fummer. The roots were accordingly collected in *Canada* with all poffible diligence; the

---

* Mr. *Ofbeck* feems to doubt whether the *Europeans* reap any advantages from the *Ginjeng* trade or not, becaufe the *Chinefe* do not value the *Canada* roots fo much as thofe of the *Chinefe-Tartary* and therefore the former bear fcarce half the price of the latter. See *Ofbeck's Voyage to China*, Vol. I. p. 223. F.

Indians

Indians especially travelled about the country in order to collect as much as they could together, and to sell it to the merchants at *Montreal*. The *Indians* in the neighbourhood of this town were likewise so much taken up with this business, that the *French* farmers were not able during that time to hire a single *Indian*, as they commonly do, to help them in the harvest. Many people feared lest by continuing for several successive years, to collect these plants without leaving one or two in each place to propagate their species, there would soon be very few of them left; which I think is very likely to happen, for by all accounts they formerly grew in abundance round *Montreal*, but at present there is not a single plant of it to be found, so effectually have they been rooted out. This obliged the *Indians* this summer to go far within the *English* boundaries to collect these roots. After the *Indians* have sold the fresh roots to the merchants, the latter must take a great deal of pains with them. They are spread on the floor to dry, which commonly requires two months and upwards, according as the season is wet or dry. During that time they must be turned once or twice every day, lest they should putrify or moulder. *Ginseng* has never been found far north

north of *Montreal*. The superior of the clergy, here and several other people, assured me that the *Chinese* value the *Canada Ginseng* as much as the *Tartarian**; and that no one ever had been entirely acquainted with the *Chinese* method of preparing it. However it is thought that amongst other preparations they dip the roots in a decoction of the leaves of *Ginseng*. The roots prepared by the *Chinese* are almost transparent, and look like horn in the inside; and the roots which are fit for use, must be heavy and compact in the inside.

The plant which throughout *Canada* bears the name of *Herba capillaris* is likewise one of those with which a great trade is carried on in *Canada*. The *English* in their plantations call it *Maiden-hair*; it grows in all their *North-American* colonies, which I travelled through, and likewise in the southern parts of *Canada*; but I never found it near *Quebec*. It grows in the woods in shady places and in a good soil †. Several people in *Albany* and *Canada*, assured me that its leaves were very much used in-

---

* This is directly opposite to Mr. *Osbeck*'s assertion. See the preceding page, 114. note †. F.

† It is the *Adiantum pedatum* of Linn. sp. pl. p. 1557. *Cornutus*, in his *Canadens. plant. historia*, p. 7. calls it *Adiantum Americanum*, and gives together with the description, a figure of it, p. 6.

stead

stead of tea, in confumptions, coughs, and all kinds of pectoral difeafes. This they have learnt from the *Indians*, who have made ufe of this plant for thefe purpofes fince times immemorial. This *American* maiden-hair is reckoned preferable in furgery to that which we have in *Europe*†; and therefore they fend a great quantity of it to *France*, every year. The price is different, and regulated according to the goodnefs of the plant, the care in preparing it, and the quantity which is to be got. For if it be brought to *Quebec* in great abundance, the price falls; and on the contrary it rifes, when the quantity gathered is but fmall. Commonly the price at *Quebec* is between five and fifteen fols a pound. The *Indians* went into the woods about this time, and travelled far above *Montreal* in queft of this plant.

The *Kitchen herbs*, fucceed very well here. The white cabbage is very fine, but fometimes fuffers greatly from worms. Onions (*Allium cepa*) are very much in ufe here, together with other fpecies of leeks. They likewife plant feveral fpecies of gourds, melons, fallads, wild fuccory or wild endive (*Cichorium Intybus*), feveral kinds of peafe, beans, *French* beans, carrots, and cucumbers. They have

† *Adiantum Capillus Veneris.* True Maiden-hair.

plenty of red beets, horseradishes and common raddishes, thyme, and marjoram. *Turneps* are sown in abundance, and used chiefly in winter. *Parsneps* are sometimes eaten, though not very common. Few people took notice of potatoes; and neither the common (*Solanum tuberosum*) nor the *Bermuda* ones (*Convolvulus Batatas*) were planted in *Canada*. When the *French* here are asked why they do not plant potatoes, they answer that they cannot find any relish in them, and they laugh at the *English* who are so fond of them. Throughout all *North-America* the root cabbage* (*Brassica gongylodes, Linn.*) is unknown to the *Swedes, English, Dutch, Irish, Germans*, and *French*. Those who have been employed in sowing and planting kitchen herbs in *Canada*, and have had some experience in gardening, told me that they were obliged to send for fresh seeds from *France* every year, because they commonly loose their strength here in the third generation, and do not produce such plants as would equal the original ones in taste and goodness.

* This is a kind of cabbage, with large round eatable roots, which grow out above the ground wherein it differs from the turnep-cabage (*Brassica Napobrassica*) whose root grows in the ground. Both are common in *Germany*, and the former likewise in *Italy*.

THE

THE *Europeans* have never been able to find any characters, much less writings, or books, among the *Indians*, who have inhabited *North-America* since time immemorial, and seem to be all of one nation, and speak the same language. These *Indians* have therefore lived in the greatest ignorance and darkness, during some centuries, and are totally unacquainted with the state of their country before the arrival of the *Europeans*, and all their knowledge of it consists in vague traditions, and mere fables. It is not certain whether any other nations possessed *America*, before the present *Indian* inhabitants came into it, or whether any other nations visited this part of the globe, before *Columbus* discovered it. It is equally unknown, whether the *Christian* religion was ever preached here in former times. I conversed with several Jesuits, who undertook long journies in this extensive country, and asked them, whether they had met with any marks that there had formerly been some *Christians* among the *Indians* which lived here? but they all answered, they had not found any. The *Indians* have ever been as ignorant of architecture and manual labour, as of science and writing. In vain does one seek for well built towns and houses, artificial

tificial fortifications, high towers and pillars, and such like, among them, which the old world can shew, from the most antient times. Their dwelling-places are wretched huts of bark, exposed on all sides to wind, and rain. All their masonry-work consists in placing a few grey rock-stones on the ground, round their fire-place, to prevent the firebrands from spreading too far in their hut, or rather to mark out the space intended for the fire-place in it. Travellers do not enjoy a tenth part of the pleasure in traversing these countries, which they must receive on their journies through our old countries, where they, almost every day, meet with some vestige or other of antiquity: now an antient celebrated town presents itself to view; here the remains of an old castle; there a field where, many centuries ago, the most powerful, and the most skilful generals, and the greatest kings, fought a bloody battle; now the native spot and residence of some great or learned man. In such places the mind is delighted in various ways, and represents all past occurrences in living colours to itself. We can enjoy none of these pleasures in *America*. The history of the country can be traced no further, than from the arrival of the *Europeans*; for every ry thing that happened before that period, is more like a fiction or a dream, than any thing that really happened. In later times there have, however, been found a few marks of antiquity, from which it may be conjectured, that *North-America* was formerly inhabited by a nation more verfed in fcience, and more civilized, than that which the *Europeans* found on their arrival here; or that a great military expedition was undertaken to this continent, from thefe known parts of the world.

THIS is confirmed by an account, which I received from Mr. *de Verandrier*, who has commanded the expedition to the fouth-fea in perfon, of which I fhall prefently give an account. I have heard it repeated by others, who have been eye-witneffes of every thing that happened on that occafion. Some years before I came into *Canada*, the then governor-general, *Chevalier de Beauharnois*, gave Mr. *de Verandrier* an order to go from *Canada*, with a number of people, on an expedition acrofs *North-America* to the fouth-fea, in order to examine, how far thofe two places are diftant from each other, and to find out, what advantages might accrue to *Canada*, or *Louifiana*, from a communication with that ocean. They fet out on horfeback from *Montreal*,

Montreal, and went as much due weſt as they could, on account of the lakes, rivers, and mountains, which fell in their way. As they came far into the country, beyond many nations, they ſometimes met with large tracts of land, free from wood, but covered with a kind of very tall graſs, for the ſpace of ſome days journey. Many of theſe fields were every where covered with furrows, as if they had been ploughed and ſown formerly. It is to be obſerved, that the nations, which now inhabit *North-America*, could not cultivate the land in this manner, becauſe they never made uſe of horſes, oxen, ploughs, or any inſtruments of huſbandry, nor had they ever ſeen a plough before the *Europeans* came to them. In two or three places, at a conſiderable diſtance from each other, our travellers met with impreſſions of the feet of grown people and children, in a rock; but this ſeems to have been no more than a *Luſus Naturæ*. When they came far to the weſt, where, to the beſt of their knowledge, no *Frenchmen*, or *European*, had ever been, they found in one place in the woods, and again on a large plain, great pillars of ſtone, leaning upon each other. The pillars conſiſted of one ſingle ſtone each, and the *Frenchmen* could not but
ſuppoſe,

suppose, that they had been erected by human hands. Sometimes they have found such stones laid upon one another, and, as it were, formed into a wall. In some of those places where they found such stones, they could not find any other sorts of stones. They have not been able to discover any characters, or writing, upon any of these stones, though they have made a very careful search after them. At last they met with a large stone, like a pillar, and in it a smaller stone was fixed, which was covered on both sides with unknown characters. This stone, which was about a foot of *French* measure in length, and between four or five inches broad, they broke loose, and carried to *Canada* with them, from whence it was sent to *France*, to the secretary of state, the count of *Maurepas*. What became of it afterwards is unknown to them, but they think it is yet preserved in his collection. Several of the Jesuits, who have seen and handled this stone in *Canada*, unanimously affirm, that the letters on it, are the same with those which in the books, containing accounts of *Tataria*, are called *Tatarian* characters *, and that, on comparing both together,

* This account seems to be highly probable, for we find
in

gether, they found them perfectly alike. Notwithstanding the questions which the *French*

in *Marco Paolo*, that *Kublai-Khan*, one of the successors of *Genghizkhan*, after the conquest of the southern part of *China*, sent ships out, to conquer the kingdom of *Japan*, or, as they call it, *Nipan-gri*, but in a terrible storm the whole fleet was cast away, and nothing was ever heard of the men in that fleet. It seems that some of these ships were cast to the shores, opposite the great *American* lakes, between forty and fifty degrees north latitude, and there probably erected these monuments, and were the ancestors of some nations, who are called *Mozemlecks*, and have some degree of civilization. Another part of this fleet, it seems, reached the country opposite *Mexico*, and there founded the *Mexican* empire, which, according to their own records, as preserved by the *Spaniards*, and in their painted annals, in *Purchas's Pilgrimage*, are very recent; so that they can scarcely remember any more than seven princes before *Motezuma* II. who was reigning when the *Spaniards* arrived there, 1519, under *Fernando Cortez*; consequently the first of these princes, supposing each had a reign of thirty-three years and four months, and adding to it the sixteen years of *Motezuma*, began to reign in the year 1270, when *Kublai-Khan*, the conqueror of all *China* and of *Japan*, was on the throne, and in whose time happened, I believe, the first abortive expedition to *Japan*, which I mentioned above, and probably furnished *North-America*, with civilized inhabitants. There is, if I am not mistaken, a great similarity between the figures of the *Mexican* idols, and those which are usual among the *Tartars*, who embrace the doctrines and religion of the *Dalaï-Lama*, whose religion *Kublai-Khan* first introduced among the *Monguls*, or *Moguls*. The savage *Indians* of *North-America*, it seems, have another origin, and are probably descended from the *Yukaghiri* and *Tchucktchi*, inhabitants of the most easterly and northerly part of *Asia*, where, according to the accounts of the *Russians*, there is but a small traject to *America*. The ferocity of these nations, similar to that of the *Americans*, their way of painting, their fondness of inebriating liquors,

(which

French on the south-sea expedition asked the people there, concerning the time when, and by whom those pillars were erected? what their traditions and sentiments concerning them were? who had wrote the characters? what was meant by them? what kind of letters they were? in what language they were written? and other circumstances; yet they could never get the least explication, the *Indians* being as ignorant of all those things, as the *French* themselves. All they could say was, that these stones had been in those places, since times immemorial. The places where the pillars stood were near nine hundred *French* miles westward of *Montreal*. The chief intention of this journey, *viz.* to come to the south-sea, and to examine its distance from *Canada*, was never attained on this occasion. For the people sent out for that purpose, were induced to take part in a war between some of the most distant *Indian*

(which the *Yukaghiri* prepare from poisonous and inebriating mushrooms, bought of the *Russians*) and many other things, show them plainly to be of the same origin. The *Eskimaux* seem to be the same nation with the inhabitants of *Greenland*, the *Samoyedes*, and *Lapponians*. *South-America*, and especially *Peru*, is probably peopled from the great unknown south continent, which is very near *America*, civilized, and full of inhabitants of various colours: who therefore might very easily be cast on the *American* continent, in boats, or proas. F.

dian nations, in which some of the *French* were taken prisoners, and the rest obliged to return. Among the last and most westerly *Indians* they were with, they heard that the south-sea was but a few days journey off; that they (the *Indians*) often traded with the *Spaniards* on that coast, and sometimes likewise they went to *Hudson's Bay*, to trade with the *English*. Some of these *Indians* had houses, which were made of earth. Many nations had never seen any *Frenchmen*; they were commonly clad in skins, but many were quite naked.

All those who had made long journies in *Canada* to the south, but chiefly westward, agreed that there were many great plains destitute of trees, where the land was furrowed, as if it had been ploughed. In what manner this happened, no one knows; for the corn-fields of a great village, or town, of the *Indians*, are scarce above four or six of our acres in extent; whereas those furrowed plains sometimes continue for several days journey, except now and then a small smooth spot, and here and there some rising grounds.

I could not hear of any more vestiges of antiquity in *Canada*, notwithstanding my careful enquiries after them. In the
con-

continuation of my journey, for the year 1750*, I shall find an opportunity of speaking of two other remarkable curiosities. Our *Swedish* Mr. *George Westmann*, A. M. has clearly, and circumstantially shewn, that our *Scandinavians*, chiefly the northern ones, long before *Columbus*'s time, have undertaken voyages to *North-America*; see his dissertation on that subject, which he read at *Abo* in 1747, for obtaining his degree.

August the 8th. THIS morning I visited the largest nunnery in *Quebec*. Men are prohibited from visiting under very heavy punishments; except in some rooms, divided by iron rails, where the men and women, that do not belong to the convent, stand without, and the nuns within the rails, and converse with each other. But to encrease the many favours which the *French* nation heaped upon me, as a *Swede*, the governor-general got the bishop's leave for me to enter the convent, and see its construction. The bishop alone has the power of granting this favour, but he does it very sparingly. The royal physician, and a surgeon, are however at liberty to go in as often as they think proper. Mr.

---

* THIS part has not yet been published.

Gaulthier, a man of great knowledge in physic and botany, was at present the royal physician here, and accompanied me to the convent. We first saw the hospital, which I shall presently describe, and then entered the convent, which forms a part of the hospital. It is a great building of stone, three stories high, divided in the inside into long galleries, on both sides of which are cells, halls, and rooms. The cells of the nuns are in the highest story, on both sides of the gallery; they are but small; not painted in the inside, but hung with paper pictures of saints, and of our Saviour on the cross. A bed with curtains, and good bed-clothes, a little narrow desk, and a chair or two, is the whole furniture of a cell. They have no fires in winter, and the nuns are forced to ly in the cold cells. On the gallery is a stove, which is heated in winter, and as all the rooms are left open, some warmth can by this means come into them. In the middle story are the rooms where they pass the day together. One of these is the room, where they are at work; this is large, finely painted and adorned, and has an iron stove. Here they were at their needle-work, embroidering, gilding, and making flowers of silk, which bear a great

simi-

similarity to the natural ones. In a word, they were all employed in such nice works, as were suitable to ladies of their rank in life. In another hall they assemble to hold their juntos. Another apartment contains those who are indisposed; but such as are more dangerously ill, have rooms to themselves. The novices, and new comers, are taught and instructed in another hall. Another is destined for their refectory, or dining-room, in which are tables on all sides; on one side of it is a small desk, on which is laid a *French* book, concerning the life of those saints who are mentioned in the New Testament. When they dine, all are silent; one of the eldest gets into the desk, and reads a part of the book before mentioned; and when they are gone through it, they read some other religious book. During the meal, they sit on that side of the table, which is turned towards the wall. Almost in every room is a gilt table, on which are placed candles, together with the picture of our Saviour on the cross, and of some saints: before these tables they say their prayers. On one side is the church, and near it a large gallery, divided from the church by rails, so that the nuns could only look into it. In this gallery they remain during divine service, and the clergyman is in the church, where the nuns reach him his sacerdotal clothes through a hole, for they are not allowed to go into the vestry, and to be in the same room with the priest. There are still several other rooms and halls here, the use of which I do not remember. The lowest story contains a kitchen, bake-house, several butteries, &c. In the garrets they keep their corn, and dry their linen. In the middle story is a balcony on the outside, almost round the whole building, where the nuns are allowed to take air. The prospect from the convent is very fine on every side; the river, the fields, and the meadows out of town, appear there to great advantage. On one side of the convent is a large garden, in which the nuns are at liberty to walk about; it belongs to the convent, and is surrounded with a high wall. There is a quantity of all sorts of fruits in it. This convent, they say, contains about fifty nuns, most of them advanced in years, scarce any being under forty years of age. At this time there were two young ladies among them, who were instructed in those things, which belong to the knowledge of nuns. They are not allowed to become nuns immediately after

after their entrance, but muſt paſs through a noviciate of two or three years, in order to try, whether they will be conſtant. For during that time it is in their power to leave the convent, if a monaſtic life does not ſuit their inclinations. But as ſoon as they are received among the nuns, and have made their vows, they are obliged to continue their whole life in it: if they appear willing to change their mode of life, they are locked up in a room, from whence they can never get out. The nuns of this convent never go further from it, than to the hoſpital, which lies near it, and even makes a part of it. They go there to attend the ſick, and to take care of them. I was told by ſeveral people here, ſome of which were ladies, that none of the nuns went into a convent, till ſhe had attained to an age in which ſhe had ſmall hopes of ever getting a huſband. The nuns of all the three convents in *Quebec* looked very old, by which it ſeems, that there is ſome foundation for this account. All agree here, that the men are much leſs numerous in *Canada*, than the women; for the men die on their voyages; many go to the *Weſt-Indies*, and either ſettle, or die, there; many are killed in battles, *&c.* Hence there seems to be a necessity of some women going into convents.

The hospital, as I have before mentioned, makes a part of the convent. It consists of two large halls, and some rooms near the apothecary's shop. In the halls are two rows of beds on each side, within each other. The beds next to the wall are furnished with curtains, the outward ones are without them. In each bed are fine bed-clothes, with clean double sheets. As soon as a sick person has left his bed, it is made again, in order to keep the hospital in cleanliness, and order. The beds are two or three yards distant, and near each is a small table. There are good iron stoves, and fine windows in this hall. The nuns attend the sick people, and bring them meat, and other necessaries. Besides them there are some men who attend, and a surgeon. The royal physician is likewise obliged to come hither, once or twice every day, look after every thing, and give prescriptions. They commonly receive sick soldiers into this hospital, who are very numerous in *July* and *August*, when the king's ships arrive, and in time of war. But at other times, when no great number of soldiers are sick, other poor people can take

take their places, as far as the number of empty beds will reach. The king finds every thing here that is requisite for the sick persons, viz. provisions, medicines, fewel, &c. Those who are very ill, are put into separate rooms, in order that the noise in the great hall may not be troublesome to them.

THE civility of the inhabitants here is more refined than that of the *Dutch* and *English*, in the settlements belonging to *Great Britain*; but the latter, on the other hand, do not idle their time away in dressing, as the *French* do here. The ladies, especially, dress and powder their hair every day, and put their locks in papers every night; which idle custom was not introduced in the *English* settlements. The gentlemen wear generally their own hair; but some have wigs. People of rank are used to wear laced cloaths, and all the crown-officers wear swords. All the gentlemen, even those of rank, the governor-general excepted, when they go into town on a day that looks likely for rain, carry their cloaks on their left arm. Acquaintances of either sex, who have not seen each other for some time, on meeting again salute with mutual kisses.

THE plants which I have collected in

Canada, and which I have partly defcribed, I pafs over as I have done before, that I may not tire the patience of my readers by a tedious enumeration. If I fhould crowd my journal with my daily botanical obfervations, and defcriptions of animals, birds, infects, ores, and the like curiofities, it would be fwelled to fix or ten times its prefent fize *. I therefore fpare all thefe things, confifting chiefly of dry defcriptions of natural curiofities, for a *Flora Canadenfis,* and other fuch like things. The fame I muft fay in regard to the obfervations I have made in phyfic. I have carefully collected all I could on this journey, concerning the medicinal ufe of the *American* plants, and the fimples, fome of which they reckon infallible †, in more than one place. But phyfic not being my principal ftudy (though from my youth I always was fond of it) I may probably have omitted remarkable circumftances in my accounts of medicines and fimples, though one cannot be too accurate in fuch cafes. The phyficians would therefore reap little or no benefit from fuch remarks, or
at

---

* What bookfeller in *Sweden* could undertake to print fuch a work at his own expence, without lofing confiderably by it?

† *Remèdes Souverains.*

at leaft they would not find them as they ought to be. This will excufe me for avoiding, as much as poffible, to mention fuch things as belong to phyfic, and are above my knowledge. Concerning the *Canada* plants, I can here add, that the further you go northward, the more you find the plants are the fame with the *Swedifh* ones: thus, on the north fide of *Quebec*, a fourth part of the plants, if not more, are the fame with the fpontaneous plants in *Sweden*. A few plants and trees, which have a particular quality, or are applied to fome particular ufe, fhall, however, be mentioned in a few words, in the fequel.

THE *Rein-deer Mofs (Lichen rangiferinus)* grows plentiful in the woods round *Quebec*. M. *Gaulthier*, and feveral other gentlemen, told me, that the *French*, on their long journies through the woods, on account of their fur trade with the *Indians*, fometimes boil this mofs, and drink the decoction, for want of better food, when their provifions are at an end; and they fay it is very nutritive. Several *Frenchmen*, who have been in the *Terra Labrador*, where there are many rein-deer (which the *French* and *Indians* here call *Cariboux)* related, that all the land there is in moft places

places covered with this rein-deer mofs, fo that the ground looks as white as fnow.

Auguſt the 10th. This day I dined with the Jefuits. A few days before, I paid my vifit to them; and the next day their prefident, and another father Jefuit, called on me, to invite me to dine with them to-day. I attended divine fervice in their church, which is a part of their houfe. It is very fine within, though it has no feats; for every one is obliged to kneel down during the fervice. Above the church is a fmall fteeple, with a clock. The building the Jefuits live in is magnificently built, and looks exceeding fine, both without and within; which gives it a fimilarity to a fine palace. It confifts of ftone, is three ftories high, exclufive of the garret, covered with flates, and built in a fquare form, like the new palace at *Stockholm*, including a large court. Its fize is fuch, that three hundred families would find room enough in it; though at prefent there were not above twenty Jefuits in it. Sometimes there is a much greater number of them, efpecially when thofe return, who have been fent as miffionaries into the country. There is a long walk along all the fides of the fquare, in every ftory, on both fides of which are either cells, halls,

or

or other apartments for the friars; and likewife their library, apothecary-fhop, &c. Every thing is very well regulated, and the Jefuits are very well accommodated here. On the outfide is their college, which is on two fides furrounded with great orchards and kitchen-gardens, in which they have fine walks. A part of the trees here, are the remains of the foreft which ftood here when the *French* began to build this town. They have befides planted a number of fruit-trees; and the garden is ftocked with all forts of plants for the ufe of the kitchen. The Jefuits dine together in a great hall. There are tables placed all round it along the walls, and feats between the tables and the walls, but not on the other fide. Near one wall is a pulpit, upon which one of the fathers gets during the meal, in order to read fome religious book; but this day it was omitted, all the time being employed in converfation. They dine very well, and their difhes are as numerous as at the greateft feafts. In this fpacious building you do not fee a fingle woman; all are fathers, or brothers; the latter of which are young men, brought up to be Jefuits. They prepare the meal, and bring it upon table; for the common fervants are not admitted.

Be-

Besides the bishop, there are three kinds of clergymen in *Canada*; viz. Jesuits, priests, and recollets. The Jesuits are, without doubt, the most considerable; therefore they commonly say here, by way of proverb, that a hatchet is sufficient to sketch out a recollet; a priest cannot be made without a chissel; but a Jesuit absolutely requires the pencil *; to shew how much one surpasses the others. The Jesuits are commonly very learned, studious, and are very civil and agreeable in company. In their whole deportment there is something pleasing; it is no wonder therefore that they captivate the minds of people. They seldom speak of religious matters; and if it happens, they generally avoid disputes. They are very ready to do any one a service; and when they see that their assistance is wanted, they hardly give one time to speak of it, falling to work immediately, to bring about what is required of them. Their conversation is very entertaining and learned, so that one cannot be tired of their company. Among all the Jesuits I have conversed with in *Canada*, I have not found one who was not possessed of these qualities in a very eminent

---

* *Pour faire un recollet il faut une hachette, pour un prêtre un ciseau, mais pour un Jesuite il faut un pinceau.*

nent degree. They have large poffeffions in this country, which the *French* king gave them. At *Montreal* they have likewife a fine church, and a little neat houfe, with a fmall but pretty garden within. They do not care to become preachers to a congregation in the town and country; but leave thefe places, together with the emoluments arifing from them, to the priefts. All their bufinefs here is to convert the heathens; and with that view their miffionaries are fcattered over every part of this country. Near every town and village, peopled by converted *Indians*, are one or two Jefuits, who take great care that they may not return to paganifm, but live as Chriftians ought to do. Thus there are Jefuits with the converted *Indians* in *Tadouffac*, *Lorette*, *Becancourt*, *St. François*, *Saut St. Louis*, and all over *Canada*. There are likewife Jefuit miffionaries with thofe who are not converted; fo that there is commonly a Jefuit in every village belonging to the *Indians*, whom he endeavours on all occafions to convert. In winter he goes on their great hunts, where he is frequently obliged to fuffer all imaginable inconveniencies; fuch as walking in the fnow all day; lying in the open air all winter; being out both in good and bad weather,

the

the *Indians* not regarding any kind of weather; lying in the *Indian* huts, which often swarm with fleas and other vermin, &c. The Jesuits undergo all these hardships for the sake of converting the *Indians*, and likewise for political reasons. The Jesuits are of great use to their king; for they are frequently able to persuade the *Indians* to break their treaty with the *English*, to make war upon them, to bring their furs to the *French*, and not to permit the *English* to come amongst them. But there is some danger attending these attempts; for when the *Indians* are in liquor, they sometimes kill the missionaries who live with them; calling them spies, or excusing themselves by saying that the brandy had killed them. These are accordingly the chief occupations of the Jesuits here. They do not go to visit the sick in the town, they do not hear the confessions, and attend at no funerals. I have never seen them go in processions in remembrance of the Virgin *Mary*, and other saints. They seldom go into a house in order to get meat; and though they be invited, they do not like to stay, except they be on a journey. Every body sees, that they are, as it were, selected from the other people, on account of their superior ge-

genius and qualities. They are here reckoned a moſt cunning ſet of people, who generally ſucceed in their undertakings, and ſurpaſs all others in acuteneſs of underſtanding. I have therefore ſeveral times obſerved that they have enemies in *Canada*. They never receive any others into their ſociety, but perſons of very promiſing parts; ſo that there are no blockheads among them. On the other hand, the prieſts receive the beſt kind of people among their order they can meet with; and in the choice of monks, they are yet leſs careful. The Jeſuits who live here, are all come from *France*; and many of them return thither again, after a ſtay of a few years here. Some (five or ſix of which are yet alive) who were born in *Canada*, went over to *France*, and were received among the Jeſuits there; but none of them ever came back to *Canada*. I know not what political reaſon hindered them. During my ſtay in *Quebec*, one of the prieſts, with the biſhop's leave, gave up his prieſthood, and became a Jeſuit. The other prieſts were very ill pleaſed with this, becauſe it ſeemed as if he looked upon their condition as too mean for himſelf. Thoſe congregations in the country that pay rents to the Jeſuits, have, however, divine ſervice

vice performed by priests, who are appointed by the bishop; and the land-rent only belongs to the Jesuits. Neither the priests nor the Jesuits carry on any trade with furs and skins, leaving that entirely to the merchants.

THIS afternoon I visited the building called the *Seminary*, where all the priests live in common. They have a great house, built of stone, with walks in it, and rooms on each side. It is several stories high, and close to it is a fine garden, full of all sorts of fruit-trees and pot-herbs, and divided by walks. The prospect from hence is the finest in *Quebec*. The priests of the seminary are not much inferior to the Jesuits in civility; and therefore I spent my time very agreeably in their company.

THE priests are the second and most numerous class of the clergy in this country; for most of the churches, both in towns and villages (the *Indian* converts excepted) are served by priests. A few of them are likewise missionaries. In *Canada* are two *seminaries*; one in *Quebec*, the other in *Montreal*. The priests of the seminary in *Montreal* are of the order of St. *Sulpitius*, and supply only the congregation on the isle of *Montreal*, and the town of the same name. At all the other churches in *Canada*,

nada, the priests belonging to the *Quebec* seminary officiate. The former, or those of the order of St. *Sulpitius,* all come from *France;* and I was assured that they never suffer a native of *Canada* to come among them. In the seminary at *Quebec,* the natives of *Canada* make the greater part. In order to fit the children of this country for orders, there are schools at *Quebec* and *St. Joachim;* where the youths are taught *Latin,* and instructed in the knowledge of those things and sciences, which have a more immediate connexion with the business they are intended for. However, they are not very nice in their choice; and people of a middling capacity are often received among them. They do not seem to have made great progress in *Latin;* for notwithstanding the service is read in that language, and they read their *Latin* Breviary, and other books, every day, yet most of them found it very difficult to speak it. All the priests in the *Quebec* seminary are consecrated by the bishop. Both the seminaries have got great revenues from the king; that in *Quebec* has above thirty thousand livres. All the country on the west side of the river St. *Lawrence,* from the town of *Quebec* to bay St. *Paul,* belongs to this seminary, besides their other possessions in the country. They lease the land to the settlers for a certain rent, which, if it be annually paid according to their agreement, the children or heirs of the settlers may remain in an undisturbed possession of the lands. A piece of land, three *arpens* * broad, and thirty, forty, or fifty *arpens* long, pays annually an *ecu* †, and a couple of chickens, or some other additional trifle. In such places as have convenient water-falls, they have built water-mills, or saw-mills, from which they annually get considerable sums. The seminary of *Montreal* possesses the whole ground on which that town stands, together with the whole isle of *Montreal*. I have been assured, that the ground-rent of the town and isle is computed at seventy thousand livres; besides what they get for saying masses, baptizing, holding confessions, attending at marriages and funerals, &c. All the revenues of ground-rent belong to the seminaries alone, and the priests in the country have no share in them. But as the seminary in *Montreal*, consisting only of sixteen priests, has greater revenues than it can expend, a large sum of money is annually sent over to France, to the chief se-

---

* A *French* acre.
† A *French* coin, value about a crown *English*.

seminary there. The land-rents belonging to the *Quebec* seminary are employed for the use of the priests in it, and for the maintenance of a number of young people, who are brought up to take orders. The priests who live in the country parishes, get the tythe from their congregation, together with the perquisites on visiting the sick, &c. In small congregations, the king gives the priests an additional sum. When a priest in the country grows old, and has done good services, he is sometimes allowed to come into the seminary in town. The seminaries are allowed to place the priests on their own estates; but the other places are in the gift of the bishop.

The recolets are the third class of clergymen in *Canada*. They have a fine large dwelling house here, and a fine church, where they officiate. Near it is a large and fine garden, which they cultivate with great application. In *Montreal*, and *Trois Rivieres*, they are lodged almost in the same manner as here. They do not endeavour to choose cunning fellows amongst them, but take all they can get. They do not torment their brains with much learning; and I have been assured, that after they have put on their monastic habit, they do not study to increase their knowledge, but forget even what little they knew before. At night they generally ly on mats, or some other hard matrasses; however, I have sometimes seen good beds in the cells of some of them. They have no possessions here, having made vows of poverty, and live chiefly on the alms which people give them. To this purpose, the young monks, or brothers, go into the houses with a bag, and beg what they want. They have no congregations in the country, but sometimes they go among the *Indians* as missionaries. In each fort, which contains forty men, the king keeps one of these monks, instead of a priest, who officiates there. The king gives him lodging, provisions, servants, and all he wants; besides two hundred livres a year. Half of it he sends to the community he belongs to; the other half he reserves for his own use. On board the king's ships are generally no other priests than these friars, who are therefore looked upon as people belonging to the king. When one of the chief priests * in the country dies, and his place cannot immediately be filled up, they send one of these friars there, to officiate whilst the place is

* *Pasteur.*

va-

vacant. Part of thefe monks come over from *France*, and part are natives of *Canada*. There are no other monks in *Canada* befides thefe, except now and then one of the order of St. *Auftin* or fome other, who comes with one of the king's fhips, but goes off with it again.

Auguft the 11th. THIS morning I took a walk out of town, with the royal phyfician M. *Gaulthier*, in order to collect plants, and to fee a nunnery at fome diftance from *Quebec*. This monaftery which is built very magnificently of ftone, lies in a pleafant fpot, furrounded with corn-fields, meadows, and woods, from whence *Quebec* and the river St. *Lawrence* may be feen; a hofpital for poor old people, cripples, &c. makes part of the monaftery, and is divided into two halls, one for men, the other for women. The nuns attend both fexes, with this difference however, that they only prepare the meal for the men and bring it in to them, give them phyfick, and take the cloth away when they have eaten, leaving the reft for male fervants. But in the hall where the women are, they do all the work that is to be done. The regulation in the hofpital was the fame as in that at *Quebec*. To fhew me a particular favour, the bifhop, at the defire of the Marquis *la Galiſſonnicre,* governor-general of *Canada,* granted me leave to fee this nunnery likewife, where no man is allowed to enter, without his leave, which is an honour he feldom confers on any body. The abbefs led me and M. *Gaulthier* through all the apartments, accompanied by a great number of nuns. Moſt of the nuns here are of noble families and one was the daughter of a governor. Many of them are old, but there are likewife fome very young ones among them, who looked very well. They feemed all to be more polite than thofe in the other nunnery. Their rooms are the fame as in the laſt place, except fome additional furniture in their cells; the beds are hung with blue curtains; there are a couple of fmall bureaux, a table between them and fome pictures on the walls. There are however no ſtoves in any cell. But thofe halls and rooms, in which they are affembled together, and in which the fick ones ly, are fupplied with an iron ſtove. The number of nuns is indeterminate here, and I faw a great number of them. Here are likewife fome probationers preparing for their reception among the nuns. A number of little girls are fent hither by their parents, to be inſtructed by the nuns in the principles of the chriſtian religion, and in

all

all forts of ladies work. The convent at a diftance looks like a palace, and, as I am told, was founded by a bifhop, who they fay is buried in a part of the church.

WE botanized till dinner-time in the neighbouring meadows, and then returned to the convent to dine with a venerable old father recolet, who officiated here as a prieft. The difhes were all prepared by nuns, and as numerous and various as on the tables of great men. There were likewife feveral forts of wine, and many preferves. The revenues of this monaftery are faid to be confiderable. At the top of the building is a fmall fteeple with a bell. Confidering the large tracts of land which the king has given in *Canada* to convents, *Jefuits,* priefts, and feveral families of rank, it feems he has very little left for himfelf.

OUR common rafp-berries, are fo plentiful here on the hills, near corn-fields, rivers and brooks, that the branches look quite red on account of the number of berries on them. They are ripe about this time, and eaten as a defert after dinner, both frefh and preferved.

THE *Mountain Afh,* or *Sorb-tree* * is pretty common in the woods hereabouts.

* *Sorbus aucuparia.*

They reckon the north-east wind the moſt piercing of all, here. Many of the beſt people here, aſſured me, that this wind when it is very violent in winter, pierces through walls of a moderate thickneſs, ſo that the whole wall on the inſide of the houſe is covered with ſnow, or a thick hoar froſt; and that a candle placed near a thinner wall is almoſt blown out by the wind which continually comes through. This wind damages the houſes which are built of ſtone, and forces the owners to repair them very frequently on the north-eaſt ſide. The north and north-eaſt winds are likewiſe reckoned very cold here. In ſummer the north wind is generally attended with rain.

The difference of climate between *Quebec* and *Montreal* is on all hands allowed to be very great. The wind and weather of *Montreal* are often entirely different from what they are at *Quebec*. The winter there is not near ſo cold as in the laſt place. Several ſorts of fine pears will grow near *Montreal*; but are far from ſucceeding at *Quebec*, where the froſt frequently kills them. *Quebec* has generally more rainy weather, ſpring begins later, and winter ſooner than at *Montreal*, where all ſorts of fruits ripen a week or two earlier than at *Quebec*.

Auguſt

August the 12th. THIS afternoon I and my servant went out of town, to stay in the country for a couple of days that I might have more leisure to examine the plants which grow in the woods here, and the state of the country. In order to proceed the better, the governor-general had sent for an *Indian* from *Lorette* to shew us the way, and teach us what use they make of the spontaneous plants hereabouts. This *Indian* was an *Englishman* by birth, taken by the *Indians* thirty years ago, when he was a boy, and adopted by them, according to their custom, instead of a relation of theirs killed by the enemy. Since that time he constantly stayed with them, became a *Roman Catholic* and married an *Indian* woman: he dresses like an *Indian*, speaks *English* and *French*, and many of the *Indian* languages. In the wars between the *French* and *English*, in this country, the *French Indians* have made many prisoners of both sexes in the *English* plantations, adopted them afterwards, and they married with people of the *Indian* nations. From hence the *Indian* blood in *Canada* is very much mixed with *European* blood, and a great part of the *Indians* now living, owe their origin to *Europe*. It is likewise remarkable, that a great part of the people they had

taken

taken during the war and incorporated with their nations, especially the young people, did not choose to return to their native country, though their parents and nearest relations came to them and endeavoured to persuade them to it, and though it was in their power to do it. The licentious life led by the *Indians*, pleased them better than that of their *European* relations; they dressed like the *Indians*, and regulated all their affairs in their way. It is therefore difficult to distinguish them, except by their colour, which is somewhat whiter than that of the *Indians*. There are likewise examples of some *Frenchmen* going amongst the *Indians* and following their way of life. There is on the contrary scarce one instance of an *Indian*'s adopting the *European* customs; but those who were taken prisoners in the war, have always endeavoured to come to their own people again, even after several years of captivity, and though they enjoyed all the privileges, that were ever possessed by the *Europeans* in *America*.

THE lands, which we passed over, were every where laid out into corn-fields, meadows, or pastures. Almost all round us the prospect presented to our view farms and farm-houses, and excellent fields and meadows. Near the town the land is

pretty

pretty flat, and interfected now and then by a clear rivulet. The roads are very good, broad, and lined with ditches on each fide, in low grounds. Further from the town, the land rifes higher and higher, and confifts as it were of terraces, one above another. This rifing ground is, however, pretty fmooth, chiefly without ftones, and covered with rich mould. Under that is the black lime-flate, which is fo common hereabouts, and is divided into fmall fhivers, and corroded by the air. Some of the ftrata were horizontal, others perpendicular; I have likewife found fuch perpendicular ftrata of lime-ftates in other places, in the neighbourhood of *Quebec*. All the hills are cultivated; and fome are adorned with fine churches, houfes, and corn-fields. The meadows are commonly in the vallies, though fome were likewife on eminencies. Soon after we had a fine profpect from one of thefe hills. *Quebec* appeared very plain to the eaftward, and the river St. *Lawrence* could likewife be feen; further diftant, on the fouth-eaft fide of that river, appears a long chain of high mountains, running generally parallel to it, though many miles diftant from it. To the weft again, at fome diftance from the rifing lands where we were, the hills changed ed into a long chain of very high mountains, lying very close to each other, and running parallel likewise to the river, that is nearly from south to north. These high mountains consist of a grey rock-stone, composed of several kinds of stone, which I shall mention in the sequel. These mountains seem to prove, that the lime-slates are of as antient a date as the grey rock-stone, and not formed in later times; for the amazing large grey rocks ly on the top of the mountains, which consist of black lime-slates.

The high meadows in *Canada* are excellent, and by far preferable to the meadows round *Philadelphia*, and in the other *English* colonies. The further I advanced northward here, the finer were the meadows, and the turf upon them was better and closer. Almost all the grass here is of two kinds, *viz.* a species of the *narrow leaved meadow grass* *; for its spikes † contain either three or four flowers; which are so exceedingly small, that the plant might easily be taken for a *bent grass* ‡; and its seeds have several small downy hairs at the bottom. The other plant, which

grows

---
* *Poa angustifolia*. Linn.
† Spiculæ tri vel quadri-floræ minimæ; femina basi pubescentia.
‡ *Agrostis*. Linn.

grow: in the meadows, is the *white clover* *. Thefe two plants form the hay in the meadows; they ftand clofe and thick together, and the meadow grafs (*poa*) is pretty tall, but has very thin ftalks. At the root of the meadow grafs, the ground was quite covered with clover, fo that one cannot wifh for finer meadows, than are found here. Almoft all the meadows have been formerly corn-fields, as appears from the furrows on the ground, which ftill remained. They can be mown but once every fummer, as fpring commences very late.

THEY were now bufied with making hay, and getting it in, and I was told, they had begun about a week ago. They have hay-ftacks near moft of their meadows, and on the wet ones, they make ufe of conic hay-ftacks. Their meadows are commonly without enclofures, the cattle being in the paftures on the other fide of the woods, and having cowherds to take care of them where they are neceffary.

THE corn-fields are pretty large. I faw no drains any where, though they feemed to be wanting in fome places. They are divided into ridges, of the breadth of two

* *Trifolium repens.* Linn. *Trifolium pratenfe album.* C. B.

or

or three yards broad, between the furrows. The perpendicular height of the middle of the ridge, from the level to the ground, is near one foot. All their corn is summer-corn; for as the cold in winter destroys the corn which lies in the ground, they never sow in autumn. I found white wheat most commonly in the fields. They have likewise large fields with pease, oats, in some places summer-rye, and now and then barley. Near almost every farm I met with cabbages, pumpions, and melons. The fields are not always sown, but ly fallow every two years. The fallow-fields are not ploughed in summer, so the weeds grow without restraint in them, and the cattle are allowed to go on them all summer *.

THE houses in the country are built promiscuously of stone, or wood. To those of stone they do not employ bricks, as there is not yet any considerable quantity of bricks made here. They therefore take what stones they can find in the neighbourhood, especially the black lime-slates. These are quite compact when broke,

* Here follows, in the original, an account of the enclosures made use of near *Quebec*, which is intended only for the *Swedes*, but not for a nation that has made such progress in agriculture and husbandry, as the *English*. F.

broke, but shiver when exposed to the air; however, this is of little consequence, as the stones stick fast in the wall, and do not fall asunder. For want of it, they sometimes make their buildings of limestone, or sand-stone, and sometimes of grey rock-stone. The walls of such houses are commonly two foot thick, and seldom thinner. The people here can have lime every where in this neighbourhood. The greater part of the houses in the country, are built of wood, and sometimes plaistered over on the outside. The chinks in the walls are filled with clay, instead of moss. The houses are seldom above one story high. In every room is either a chimney or stove, or both together. The stoves have the form of an oblong square; some are entirely of iron, about two feet and a half long, one foot and a half, or two feet, high, and near a foot and a half broad; these iron stoves are all cast at the iron-works at *Trois Rivieres*. Some are made of bricks, or stones, not much larger than the iron stoves, but covered at top with an iron plate. The smoke from the stoves is conveyed up the chimney, by an iron pipe. In summer the stoves are removed.

THIS

This evening we arrived at *Lorette*, where we lodged with the Jesuits.

August the 13th. In the morning we continued our journey through the woods to the high mountains, in order to see what scarce plants and curiosities we could get there. The ground was flat at first, and covered with a thick wood all round, except in marshy places. Near half the plants, which are to be met with here, grow in the woods and morasses of *Sweden*.

We saw wild Cherry-trees here, of two kinds, which are probably mere varieties, though they differ in several respects. Both are pretty common in *Canada*, and both have red berries. One kind, which is called *Cerisier* by the *French*, tastes like our *Alpine* cherries, and their acid contracts the mouth, and cheeks. The berries of the other sort have an agreeable sourness, and a pleasant taste *.

The three-leaved Hellebore † grows in great plenty in the woods, and in many places it covers the ground by itself. However, it commonly chooses mossy places,

that

---

* The kind called *Cerisier* by the *French*, I described thus in my journal: *Cerasus foliis ovatis serratis, serraturis profundis fere subulatis, fructu racemoso.* The other thus: *Cerasus foliis lanceolatis, crenato-serratis, acutis, fructu fere solitario.*

† Helleborus trifolius.

that are not very wet; and the wood-forrel*, with the *Mountain Enchanter's Nightshade*†, are its companions. Its feeds were not yet ripe, and moſt of the ſtalks had no feeds at all. This plant is called *Tiſſavoyanne jaune* by the *French*, all over *Canada*. Its leaves and ſtalks are uſed by the *Indians*, for giving a fine yellow colour to ſeveral kinds of work, which they make of prepared ſkins. The *French*, who have learnt this from them, dye wool and other things yellow with this plant.

WE climbed with a great deal of difficulty to the top of one of the higheſt mountains here, and I was vexed to find nothing at its ſummit, but what I had ſeen in other parts of *Canada* before. We had not even the pleaſure of a proſpect, becauſe the trees, with which the mountain is covered, obſtructed it. The trees that grow here are a kind of hornbeam, or *Carpinus Oſtrya*, Linn. the *American* elm, the red maple, the ſugar-maple, that kind of maple which cures ſcorched wounds (which I have not yet deſcribed), the beech, the common birch-tree, the ſugar-birch ‡, the ſorb-tree, the *Canada*

---

* *Oxalis Acetoſella*, Linn.
† *Circæa alpina*, Linn.
‡ *Betula nigra*, Linn.

pine, called *Peruſſe*, the mealy-tree with dentated leaves *, the aſh, the cherry-tree, (*Ceriſier*) juſt before deſcribed, and the berry-bearing yew.

The Gnats in this wood were more numerous than we could have wiſhed. Their bite cauſed a bliſtering of the ſkin; and the Jeſuits at *Lorette* ſaid, the beſt preſervative againſt their attacks is to rub the face, and naked parts of the body, with greaſe. Cold water they reckon the beſt remedy againſt the bite, when the wounded places are waſhed with it, immediately after.

At night we returned to *Lorette*, having accurately examined the plants of note we met with to-day.

Auguſt the 14th. *Lorette* is a village, three *French* miles to the weſtward of *Quebec*. Inhabited chiefly by *Indians* of the *Huron* nation, converted to the Roman catholic religion. The village lies near a little river, which falls over a rock there, with a great noiſe, and turns a ſaw-mill, and a flour-mill. When the Jeſuit, who is now with them, arrived among them, they lived in their uſual huts, which are made like thoſe of the *Laplanders*. They

---

* *Viburnum dentatum*, Linn.

have since laid aside this custom, and built all their houses after the *French* fashion. In each house are two rooms, *viz.* their bed-room, and the kitchen on the outside before it. In the room is a small oven of stone, covered at top with an iron plate. Their beds are near the wall, and they put no other clothes on them, than those which they are dressed in. Their other furniture and utensils, look equally wretched. Here is a fine little church, with a steeple and bell. The steeple is raised pretty high, and covered with white tin plates. They pretend, that there is some similarity between this church in its figure and disposition, and the *Santa Casa*, at *Loretto* in *Italy*, from whence this village has got its name. Close to the church is a house built of stone, for the clergymen, who are two Jesuits, that constantly live here. The divine service is as regularly attended here, as in any other Roman catholic church; and I was pleased with seeing the alacrity of the *Indians*, especially of the women, and hearing their good voices, when they sing all sorts of hymns in their own language. The *Indians* dress chiefly like the other adjacent *Indian* nations; the men, however, like to wear waistcoats, or jackets, like the *French*. The women keep exactly to the *Indian* drefs. It is certain, that thefe *Indians* and their anceftors, long fince, on being converted to the *Chriftian* religion, have made a vow to God, never to drink ftrong liquors. This vow they have kept pretty inviolable hitherto, fo that one feldom fees one of them drunk, though brandy and other ftrong liquors are goods, which other *Indians* would fooner be killed for, than part with them.

These *Indians* have made the *French* their patterns in feveral things, befides the houfes. They all plant maize; and fome have fmall fields of wheat, and rye. Many of them keep cows. They plant our common fun-flower* in their maize-fields, and mix the feeds of it into their *fagamite,* or maize-foup. The maize, which they plant here, is of the fmall fort, which ripens fooner than the other: its grains are fmaller, but give more and better flour in proportion. It commonly ripens here at the middle, fometimes however, at the end of *Auguft.*

The *Swedifh* winter-wheat, and winter-rye, has been tried in *Canada,* to fee how well it would fucceed; for they employ nothing but fummer-corn here, it having

* Helianthus annuus.

been

been found, that the *French* wheat and rye dies here in winter, if it be sown in autumn. Dr. *Sarrazin* has therefore (as I was told by the eldest of the two Jesuits here) got a small quantity of wheat and rye, of the winter-corn sort from *Sweden*. It was sown in autumn, not hurt by the winter, and bore fine corn. The ears were not so large as those of the *Canada* corn, but weighed near twice as much, and gave a greater quantity of finer flour, than that summer-corn. Nobody could tell me, why the experiments have not been continued. They cannot, I am told, bake such white bread here, of the summer-corn, as they can in *France*, of their winter-wheat. Many people have assured me, that all the summer-corn, now employed here, came from *Sweden*, or *Norway*: for the *French*, on their arrival, found the winters in *Canada* too severe for the *French* winter-corn, and their summer-corn did not always ripen, on account of the shortness of summer. Therefore they began to look upon *Canada*, as little better than an useless country, where nobody could live; till they fell upon the expedient of getting their summer-corn from the most northern parts of *Europe*, which has succeeded very well.

THIS day I returned to *Quebec*, making botanical obfervations by the way.

Auguſt the 15th. THE new governor-general of all *Canada*, the marquis *de la Jonquiere*, arrived laſt night in the river before *Quebec*; but it being late, he reſerved his public entrance for to-day. He had left *France* on the ſecond of *June*, but could not reach *Quebec* before this time, on account of the difficulty which great ſhips find in paſſing the ſands in the river St. *Lawrence*. The ſhips cannot venture to go up, without a fair wind, being forced to run in many bendings, and frequently in a very narrow channel. To-day was another great feaſt, on account of the Aſcenſion of the Virgin *Mary*, which is very highly celebrated in Roman catholic countries. This day was accordingly doubly remarkable, both on account of the holiday, and of the arrival of the new governor-general, who is always received with great pomp, as he repreſents a vice-roy here.

ABOUT eight o'clock the chief people in town aſſembled at the houſe of Mr. *de Vaudreuil*, who had lately been nominated governor of *Trois Rivieres*, and lived in the lower town, and whoſe father had likewiſe been governor-general of *Canada*. Thither came likewiſe the marquis *de la Galiſſonniere*,

Galiſſonniere, who had till now been governor-general, and was to ſail for *France*, with the firſt opportunity. He was accompanied by all the people belonging to the government. I was likewiſe invited to ſee this feſtivity. At half an hour after eight the new governor-general went from the ſhip into a barge, covered with red cloth, upon which a ſignal with cannons was given from the ramparts, for all the bells in the town to be ſet a-ringing. All the people of diſtinction went down to the ſhore to ſalute the governor, who, on alighting from the barge, was received by the marquis *la Galiſſonniere*. After they had ſaluted each other, the commandant of the town addreſſed the new governor-general in a very elegant ſpeech, which he anſwered very conciſely; after which all the cannon on the ramparts gave a general ſalute. The whole ſtreet, up to the cathedral, was lined with men in arms, chiefly drawn out from among the burgheſſes. The governor-general then walked towards the cathedral, dreſſed in a ſuit of red, with abundance of gold lace. His ſervants went before him in green, carrying fire-arms on their ſhoulders. On his arrival at the cathedral, he was received by the biſhop of *Canada*, and the whole clergy aſſembled. The biſhop was arrayed in in his pontifical robes, and had a long gilt tiara on his head, and a great crozier of maffy filver in his hand. After the bifhop had addreffed a fhort fpeech to the governor-general, a prieft brought a filver crucifix on a long ftick, (two priefts with lighted tapers in their hands, going on each fide of it) to be kiffed by the governor. The bifhop and the priefts then went through the long walk, up to the choir. The fervants of the governor-general followed with their hats on, and arms on their fhoulders. At laft came the governor-general and his fuite, and after them a croud of people. At the beginning of the choir the governor-general, and the general *de la Galiffonniere*, ftopt before a chair covered with red cloth, and ftood there during the whole time of the celebration of the mafs, which was celebrated by the bifhop himfelf. From the church he went to the palace, when the gentlemen of note in the town, afterwards went to pay their refpects to him. The religious of the different orders, with their refpective fuperiors, likewife came to him, to teftify their joy on account of his happy arrival. Among the numbers that came to vifit him, none ftaid to dine, but thofe that were invited beforehand, among which I had the honour hour to be. The entertainment lasted very long, and was as elegant as the occasion required.

The governor-general, marquis *de la Jonquiere*, was very tall, and at that time something above sixty years old. He had fought a desperate naval battle with the *English* in the last war, but had been obliged to surrender, the *English* being, as it was told, vastly superior in the number of ships and men. On this occasion he was wounded by a ball, which entered one side of his shoulder, and came out at the other. He was very complaisant, but knew how to preserve his dignity, when he distributed favours.

Many of the gentlemen, present at this entertainment, asserted that the following expedient had been successfully employed to keep wine, beer, or water, cool during summer. The wine, or other liquor, is bottled; the bottles are well corked, hung up into the air, and wrapped in wet clouts. This cools the wine in the bottles, notwithstanding it was quite warm before. After a little while the clouts are again made wet, with the coldest water that is to be had, and this is always continued. The wine, or other liquor, in the bottles is then always colder, than the water with
which

which the clouts are made wet. And though the bottles should be hung up in the sunshine, the above way of proceeding will always have the same effect *.

August the 16th. THE occidental *Arbor vitæ* †, is a tree which grows very plentiful in *Canada*, but not much further south. The most southerly place I have seen it in, is a place a little on the south side of *Saratoga*, in the province of *New-York*, and likewise near *Casses*, in the same province, which places are in forty-two degrees and ten minutes north latitude.

Mr. *Bartram*, however, informed me, that he had found a single tree of this kind in *Virginia*, near *the falls* in the river *James*. Doctor *Colden* likewise asserted, that he had seen it in many places round his seat *Coldingham*,

---

* It has been observed by several experiments, that any liquor dipt into another liquor, and then exposed into the air for evaporation, will get a remarkable degree of cold; the quicker the evaporation succeeds, after repeated dippings, the greater is the cold. Therefore spirit of wine evaporating quicker than water, cools more than water; and spirit of sal ammoniac, made by quick-lime, being still more volatile than spirit of wine, its cooling quality is still greater. The evaporation succeeds better by moving the vessel containing the liquor, by exposing it to the air, and by blowing upon it, or using a pair of bellows. See *de Mairan, Dissertation sur la Glace, Prof. Richman* in *Nov. Comment. Petrop.* ad an. 1747, & 1748. p. 284. and Dr. *Cullen* in the *Edinburgh physical and literary Essays and Observations.* Vol. II. p. 145. F.

† *Thuja occidentalis*, Linn.

ingham, which lies between *New-York*, and *Albany*, about forty-one degrees thirty minutes north latitude. The *French*, all over *Canada*, call it *Cedre blanc*. The *English* and *Dutch* in *Albany*, likewise call it the white Cedar. The *English* in *Virginia*, have called a Thuya, which grows with them, a *Juniper*.

THE places and the soil where it grows best, are not always alike, however it generally succeeds in such ground where its roots have sufficient moisture. It seems to prefer swamps, marshes, and other wet places to all others, and there it grows pretty tall. Stony hills, and places where a number of stones ly together, covered with several kinds of mosses*, seemed to be the next in order where it grows. When the sea shores were hilly, and covered with mossy stones, the Thuya seldom failed to grow on them. It is likewise seen now and then on the hills near rivers, and other high grounds, which are covered with a dust like earth or mould; but it is to be observed that such places commonly carry a sourish water with them, or receive moisture from the upper countries. I have however seen it growing in some pretty dry places; but there it never

---

* *Lichen, Bryum, Hypnum.*

comes

comes to any confiderable fize. It is pretty frequent in the clefts of mountains, but cannot grow to any remarkable height or thicknefs. The talleft trees. I have found in the woods in *Canada*, were about thirty or thirty-fix feet high. A tree of exactly ten inches diameter had ninety-two rings round the ftem*; another of one foot and two inches in diameter had one hundred and forty-two rings †.

THE inhabitants of *Canada* generally make ufe of this tree in the following cafes. It being reckoned the moft durable wood in *Canada*, and which beft withftands putrefaction, fo as to remain undamaged for above a man's age, enclofures of all kinds are fcarce made of any other than this wood. all the pofts which are driven into the ground, are made of the Thuya wood. The palifades round the forts in *Canada* are likewife made of the fame wood. The planks in the houfes are made of it; and the thin narrow pieces of wood which form both the ribs and the bottom of the barkboats, commonly made ufe of here, are taken from this wood, becaufe it is pliant

---

* Of thefe rings or circles, it is well known all trees get t one every year, fo that they ferve to afcertain the age of trees, and the quicknefs, or flownefs of its growth. F.
The bark is not included, when I fpeak of the diameters of trees.

enough

enough for the purpose, especially whilst it is fresh, and likewise because it is very light. The Thuya wood is reckoned one of the best for the use of lime-kilns. Its branches are used all over *Canada* for besoms; and the twigs and leaves of it being naturally bent together, seem to be very proper for the purpose. The *Indians* make such besoms and bring them to the towns for sale, nor do I remember having seen any besoms of any other wood. The fresh branches have a peculiar, agreeable scent, which is pretty strongly smelled in houses where they make use of besoms of this kind.

THIS Thuya is made use of for several medicinal purposes. The commandant of Fort St. *Frederic*, M. *de Lusignan*, could never sufficiently praise its excellence for rheumatic pains. He told me he had often seen it tried, with remarkable good success, upon several persons, in the following manner. The fresh leaves are pounded in a mortar, and mixed with hog's grease, or any other grease. This is boiled together till it becomes a salve, which is spread on linen, and applied to the part where the pain is. The salve gives certain relief in a short time. Against violent pains, which move up and down in the thighs, and sometimes spread all over the body, they recommend

mend the following remedy. Take of the leaves of a kind of *Polypody* * four-fifths, and of the cones of the Thuya one-fifth, both reduced to a coarse powder by themselves, and mixed together afterwards. Then pour milk-warm water on it, so as to make a poultice, which spread on linen, and wrap it round the body: but as the poultice burns like fire, they commonly lay a cloth between it and the body, otherwise it would burn and scorch the skin. I have heard this remedy praised beyond measure, by people who said they had experienced its good effects. An *Iroquese Indian* told me, that a decoction of Thuya leaves was used as a remedy for the cough. In the neighbourhood of *Saratoga*, they use this decoction in the intermitting fevers.

The Thuya tree keeps its leaves, and is green all winter. Its seeds are ripe towards the end of *September*, old style. The fourth of *October* of this year, 1749, some of the cones, especially those which stood much exposed to the heat of the sun, had already dropt their seeds, and all the other cones were opening in order to shed them. This tree has, in common with many other *Ame-*

---

* *Polypedium fronde pinnata, pinnis alternis ad basin superne appendiculatis.*

rican

rican trees, the quality of growing plentiful in marshes and thick woods, which may be with certainty called its native places. However, there is scarce a single Thuya tree in those places which bears seeds; if, on the other hand, a tree accidentally stands on the outside of a wood, on the sea shore, or in a field, where the air can freely come at it, it is always full of seeds. I have found this to be the case with the Thuya, on innumerable occasions. It is the same likewise with the sugar-maple, the maple which is good for healing scorched wounds, the white fir-tree, the pine called *Peruffe*, the mulberry-tree and several others.

August the 17th. This day I went to see the nunnery of the *Ursulines*, which is disposed nearly in the same way as the two other nunneries. It lies in the town and has a very fine church. The nuns are renowned for their piety, and they go less abroad than any others. The men are likewise not allowed to go into this monastery, but by the special licence of the bishop, which is given as a great favour; the royal physician, and the surgeon are alone entitled to go in as often as they please, to visit the sick. At the desire of the marquis *de la Galissonniere* the bishop granted me leave to visit this monastery together with the royal physician

physician Mr. *Gaulthier*. On our arrival we were received by the abbess, who was attended by a great number of nuns, for the most part old ones. We saw the church; and, it being *Sunday*, we found some nuns on every side of it kneeling by themselves and saying prayers. As soon as we came into the church, the abbess and the nuns with her dropt on their knees, and so did M. *Gaulthier* and myself. We then went to an apartment or small chapel dedicated to the *Virgin Mary*, at the entrance of which, they all fell on their knees again. We afterwards saw the kitchen, the dining hall and the apartment they work in, which is large and fine. They do all sorts of neat work there, gild pictures, make artificial flowers, &c. The dining hall is disposed in the same manner as in the other two monasteries. Under the tables are small drawers for each nun to keep her napkin, knife and fork, and other things in. Their cells are small, and each nun has one to herself. The walls are not painted; a little bed, a table with a drawer, and a crucifix, and pictures of saints on it, and a chair, constitute the whole furniture of a cell. We were then led into a room full of young ladies about twelve years old and below that age, sent hither by their parents to be in-

structed

structed in reading, and in matters of religion. They are allowed to go to visit their relations once a day, but must not stay away long. When they have learnt reading, and have received instructions in religion, they return to their parents again. Near the monastery, is a fine garden, which is surrounded with a high wall. It belongs to this institution, and is stocked with all sorts of kitchen-herbs and fruit-trees, When the nuns are at work, or during dinner, every thing is silent in the rooms, unless some one of them reads to the others; but after dinner, they have leave to take a walk for an hour or two in the garden, or to divert themselves within-doors. After we had seen every thing remarkable here, we took our leave, and departed.

ABOUT a quarter of a *Swedish* mile to the west of *Quebec*, is a well of mineral waters, which carries a deal of iron ocker with it, and has a pretty strong taste. M. *Gaulthier* said, that he had prescribed it with success in costive cases and the like diseases.

I have been assured, that there are no snakes in the woods and fields round *Quebec*, whose bite is poisonous; so that one can safely walk in the grass. I have never found any that endeavoured to bite, and all were very fearful. In the south parts of *Canada*, it is not adviseable to be off one's guard.

A very small species of black ants * live in ant-hills, in high grounds, in woods; they look exactly like our *Swedish* ants, but are much less.

August the 21st. To-day there were some people of three *Indian* nations in this country with the governor-general, viz. *Hurons, Mickmacks*, and *Anies* †; the last of which are a nation of *Iroquese*, and allies of the *English*, and were taken prisoners in the last war.

The *Hurons* are some of the same *Indians* with those who live at *Lorette*, and have received the christian religion. They are tall, robust people, well shaped, and of a copper colour. They have short black hair, which is shaved on the forehead, from one ear to the other. None of them wear hats or caps. Some have ear-rings, others not. Many of them have the face painted all over with vermillion; others have only strokes of it on the forehead, and near the ears; and some paint their hair with vermillion. Red is the colour they chiefly make use of in painting themselves; but I

* *Formica nigra.* Linn.
† Probably *Onidoes*.

have

have likewife feen fome, who had daubed their face with a black colour. Many of them have figures in the face, and on the whole body, which are ftained into the fkin, fo as to be indelible. The manner of making them fhall be defcribed in the fequel. Thefe figures are commonly black; fome have a fnake painted in each cheek, fome have feveral croffes, fome an arrow, others the fun, or any thing elfe their imagination leads them to. They have fuch figures likewife on the breaft, thighs, and other parts of the body; but fome have no figures at all. They wear a fhirt, which is either white or checked, and a fhaggy piece of cloth, which is either blue or white, with a blue or red ftripe below. This they always carry over their fhoulders, or let it hang down, in which cafe they wrap it round their middle. Round their neck, they have a ftring of violet wampums, with little white wampums between them. Thefe wampums are fmall, of the figure of oblong pearls, and made of the fhells which the *Englifh* call clams*. I fhall make a more particular mention of them in the fequel. At the end of the wampum ftrings, many of the *Indians* wear a large

* *Venus mercenaria.* Linn.

French filver coin, with the king's effigy, on their breafts. Others have a large fhell on the breaft, of a fine white colour, which they value very high, and is very dear; others, again, have no ornament at all round the neck. They all have their breafts uncovered. Before them hangs their tobacco-pouch, made of the fkin of an animal, and the hairy fide turned outwards. Their fhoes are made of fkins, and bear a great refemblance to the fhoes without heels, which the women in *Finland* make ufe of. Inftead of ftockings, they wrap the legs in pieces of blue cloth, as I have feen the *Ruffian* boors do.

The *Mickmacks* are dreffed like the *Hurons*, but diftinguifh themfelves by their long ftrait hair, of a jetty-black colour. Almoft all the *Indians* have black ftrait hair; however, I have met with a few, whofe hair was pretty much curled. But it is to be obferved, that it is difficult to judge of the true complexion of the *Canada Indians*, their blood being mixed with the *Europeans*, either by the adopted prifoners of both fexes, or by the *Frenchmen*, who travel in the country, and often contribute their fhare towards the encreafe of the *Indian* families, their women not being very fhy. The *Mickmacks* are commonly

not

not so tall as the *Hurons*. I have not seen any *Indians* whose hair was as long and strait as theirs. Their language is different from that of the *Hurons*; therefore there is an interpreter here for them on purpose.

The *Anies* are the third kind of *Indians* which came hither. Fifty of them went out in the war, being allies of the *English*, in order to plunder in the neighbourhood of *Montreal*. But the *French*, being informed of their scheme, laid an ambush, and killed with the first discharge of their guns forty-four of them; so that only the four who were here to-day saved their lives, and two others, who were ill at this time. They are as tall as the *Hurons*, whose language they speak. The *Hurons* seem to have a longer, and the *Anies* a rounder face. The *Anies* have something cruel in their looks; but their dress is the same as that of the other *Indians*. They wear an oblong piece of white tin between the hair which lies on the neck. One of those I saw had taken a flower of the rose mallow, out of a garden, where it was in full blossom at this time, and put it among the hair at the top of his head. Each of the *Indians* has a tobacco-pipe of grey lime-stone, which is blackened afterwards, and has a long tube of wood. There were no *Indian* women present at this

this enterview. As foon as the governor-general came in, and was feated in order to fpeak with them, the *Mickmacks* fat down on the ground, like *Laplanders*, but the other *Indians* took chairs.

THERE is no printing-prefs in *Canada*, tho' there formerly was one: but all books are brought from *France*, and all the orders made in the country are written, which extends even to the paper-currency. They pretend that the prefs is not yet introduced here, left it fhould be the means of propagating libels againft the government, and religion. But the true reafon feems to ly in the poornefs of the country, as no printer could put off a fufficient number of books for his fubfiftence; and another reafon may be, that *France* may have the profit arifing from the exportation of books hither.

THE meals here are in many refpects different from thofe in the *Englifh* provinces. This perhaps depends upon the difference of cuftom, tafte, and religion, between the two nations. They eat three meals a day, *viz.* breakfaft, dinner, and fupper. They breakfaft commonly between feven and eight. For the *French* here rife very early, and the governor-general can be fpoke to at feven o'clock, which

which is the time when he has his levee.
Some of the men dip a piece of bread in
brandy, and eat it; others take a dram of
brandy, and eat a piece of bread after it.
Chocolate is likewife very common for breakfaft, and many of the ladies drink coffee.
Some eat no breakfaft at all. I have never
feen tea made ufe of; perhaps becaufe they
can get coffee and chocolate from the
French provinces in *South-America*; but
muft get tea from *China*, for which it is
not worth their while to fend the money
out of their country. Dinner is pretty
exactly at noon. People of quality have
a great variety of difhes, and the reft follow their example, when they invite ftrangers. The loaves are oval, and baked of
wheat flour. For each perfon they put a
plate, napkin, fpoon, and fork. Sometimes they likewife give knives; but they
are generally omitted, all the ladies and
gentlemen being provided with their own
knives. The fpoons and forks are of filver, and the plates of *Delft* ware. The
meal begins with a foup, with a good deal
of bread in it. Then follow frefh meats
of various kinds, boiled, and roafted, poultry, or game, fricaffees, ragoos, *&c.* of
feveral forts; together with different kinds
of fallads. They commonly drink red
claret

claret at dinner, mixed with water; and spruce beer is likewise much in use. The ladies drink water, and sometimes wine. After dinner the fruit and sweet-meats are served up, which are of many different kinds, *viz.* walnuts from *France*, or *Canada*, either ripe, or pickled; almonds, raisins, haselnuts, several kinds of berries, which are ripe in the summer season, such as currants, cran-berries, which are preserved in treacle; many preserves in sugar as straw-berries, rasp-berries, black-berries, and moss-berries. Cheese is likewise a part of the desert, and so is milk, which they eat last of all with sugar. Friday and Saturday they eat no flesh, according to the Roman catholic rites; but they well know how to guard against hunger. On those days they boil all sorts of kitchen-herbs, and fruit; fishes, eggs, and milk, prepared in various ways. They cut cucumbers into slices, and eat them with cream, which is a very good dish. Sometimes they put whole cucumbers on the table, and every body that likes them takes one, peels, and slices it, and dips the slices into salt, eating them like raddishes. Melons abound here, and are always eaten with sugar. They never put any sugar into wine, or brandy, and upon the whole, they and the

English

English do not use half so much sugar, as we do in *Sweden*; though both nations have large sugar-plantations in their *West-Indian* possessions. They say no grace before, or after their meals, but only cross themselves, which is likewise omitted by some. Immediately after dinner, they drink a dish of coffee, without cream. Supper is commonly at seven o'clock, or between seven and eight at night, and the dishes the same as at dinner. Pudding and punch is not to be met with here, though the latter is well known.

August the 23d. IN many places hereabouts they use their dogs to fetch water out of the river. I saw two great dogs to day put before a little cart, one before the other. They had neat harness, like horses, and bits in their mouths. In the cart was a barrel. The dogs are directed by a boy, who runs behind the cart, and as soon as they come to the river, they jump in, of their own accord. When the barrel is filled, the dogs draw their burthen up the hill again, to the house they belong to. I have frequently seen dogs employed in this manner, during my stay at *Quebec*. Sometimes they put but one dog before the water-carts, which are made small on purpose. The dogs are not very great, hardly ly of the fize of our common farmers dogs. The boys that attend them have great whips, with which they make them go on occafionally. I have feen them fetch not only water, but likewife wood, and other things. In winter it is cuftomary in *Canada*, for travellers to put dogs before little fledges, made on purpofe to hold their clothes, provifions, &c. Poor people commonly employ them on their winter-journies, and go on foot themfelves. Almoft all the wood, which the poorer people in this country fetch out of the woods in winter, is carried by dogs, which have therefore got the name of horfes of the poor people. They commonly place a pair of dogs before each load of wood. I have likewife feen fome neat little fledges, for ladies to ride in, in winter; they are drawn by a pair of dogs, and go fafter on a good road, than one would think. A middle-fized dog is fufficient to draw a fingle perfon, when the roads are good. I have been told by old people, that horfes were very fcarce here in their youth, and almoft all the land-carriage was then effected by dogs. Several *Frenchmen*, who have been among the *Efquimaux* on *Terra Labrador*, have affured me, that they not only make ufe of dogs for drawing drays, with their provifions, provisions, and other neceffaries, but are likewife drawn by them themfelves, in little fledges.

Auguft the 25th. THE high hills, to the weft of the town, abound with fprings. Thefe hills confift of the black lime-flate, before mentioned, and are pretty fteep, fo that it is difficult to get to the top. Their perpendicular height is about twenty or four and twenty yards. Their fummits are deftitute of trees, and covered with a thin cruft of earth, lying on the lime-flates, and are employed for corn-fields, or paftures. It feems inconceivable therefore, from whence thefe naked hills could take fo many running fprings, which in fome places gufh out of the hills, like torrents. Have thefe hills the quality of attracting the water out of the air in the day time, or at night? Or are the lime-flates more apt to it, than others?

ALL the horfes in *Canada* are ftrong, well made, fwift, as tall as the horfes of our cavalry, and of a breed imported from *France*. The inhabitants have the cuftom of docking the tails of their horfes, which is rather hard upon them here, as they cannot defend themfelves againft the numerous fwarms of gnats, gad-flies, and horfe-flies. They put the horfes one before

fore the other in their carts, which has probably occasioned the docking of their tails, as the horses would hurt the eyes of those behind them, by moving their tails backwards and forwards. The governor-general, and a few of the chief people in town, have coaches, the rest make use of open horse-chairs. It is a general complaint, that the country people begin to keep too many horses, by which means the cows are kept short of food in winter.

The cows have likewise been imported from *France*, and are of the size of our common *Swedish* cows. Every body agreed that the cattle, which were born of the original *French* breed, never grow up to the same size. This they ascribe to the cold winters, during which they are obliged to put their cattle into stables, and give them but little food. Almost all the cows have horns, a few, however, I have seen without them. A cow without horns would be reckoned an unheard of curiosity in *Pensylvania*. Is not this to be attributed to the cold? The cows give as much milk here as in *France*. The beef and veal at *Quebec*, is reckoned fatter and more palatable than at *Montreal*. Some look upon the salty pastures below *Quebec*, as the cause of this difference. But this does

does not seem sufficient; for most of the cattle, which are sold at *Quebec*, have no meadows with *Arrow-headed grass**, on which they graze. In *Canada* the oxen draw with the horns, but in the *English* colonies they draw with their *withers*, as horses do. The cows vary in colour; however, most of them are either red, or black.

Every countryman commonly keeps a few sheep, which supply him with as much wool as he wants to cloth himself with. The better sort of clothes are brought from *France*. The sheep degenerate here, after they are brought from *France*, and their progeny still more so. The want of food in winter is said to cause this degeneration.

I have not seen any goats in *Canada*, and I have been assured that there are none. I have seen but very few in the *English* colonies, and only in their towns, where they are kept on account of some sick people, who drink the milk by the advice of their physicians.

The harrows are triangular; two of the sides are six feet, and the third four feet long. The teeth, and every other part of the harrows are of wood. The teeth are

* *Triglochin.*

about

about five inches long, and about as much diſtant from each other.

The proſpect of the country about a quarter of a mile *Swediſh*, north of *Quebec*, on the weſt ſide of the river St. *Lawrence*, is very fine. The country is very ſteep towards the river, and grows higher as you go further from the water. In many places it is naturally divided into terraces. From the heights, one can look a great way: *Quebec* appears very plain to the ſouth, and the river St. *Lawrence* to the eaſt, on which were veſſels ſailing up and down. To the weſt are the high mountains, which the hills of the river end with. All the country is laid out for corn-fields, meadows, and paſtures; moſt of the fields were ſown with wheat, many with white oats, and ſome with peaſe. Several fine houſes and farms are interſperſed all over the country, and none are ever together. The dwelling-houſe is commonly built of black lime-ſlates, and generally white-waſhed on the outſide. Many rivulets and brooks roll down the high grounds, above which the great mountains ly, and which conſiſt entirely of the black lime-ſlates, that ſhiver in pieces in the open air. On the lime-ſlates lies a mould of two or three feet in depth. The ſoil in the corn-fields is always mixed with little

little pieces of the lime-flate. All the rivulets cut their beds deep into the ground; fo that their fhores are commonly of lime-flate. A dark-grey lime-ftone is fometimes found among the ftrata, which, when broke, fmells like ftink-ftone.

THEY were now building feveral fhips below *Quebec*, for the king's account. However, before my departure, an order arrived from *France*, prohibiting the further building of fhips of war, except thofe which were already on the ftocks; becaufe they had found, that the fhips built of *American* oak do not laft fo long as thofe of *European* oak. Near *Quebec* is found very little oak, and what grows there is not fit for ufe, being very fmall; therefore they are obliged to fetch their oak timber from thofe parts of *Canada* which border upon *New-England*. But all the *North-American* oaks have the quality of lafting longer, and withftanding putrefaction better, the further north they grow, and *vice verfâ*. The timber from the confines of *New-England* is brought in floats or rafts on the rivers near thofe parts, and near the lake St. *Pierre*, which fall into the great river St. *Lawrence*. Some oak is likewife brought from the country between *Montreal* and Fort St. *Frederic*, or Fort *Champlain*; but it it is not reckoned so good as the first, and the place it comes from is further distant.

August the 26th. THEY shewed a green earth, which had been brought to the general, marquis *de la Galissonniere*, from the upper parts of *Canada*. It was a clay, which cohered very fast together, and was of a green colour throughout, like verdigrease.†

ALL the brooks in *Canada* contain crawfish, of the same kind with ours. The *French* are fond of eating them, and say they are vastly decreased in number since they have begun to catch them.

THE common people in the country, seem to be very poor. They have the necessaries of life, and but little else. They are content with meals of dry bread and water, bringing all other provisions, such as butter, cheese, flesh, poultry, eggs, &c. to town, in order to get money for them, for which they buy clothes and brandy for themselves, and dresses for their women. Notwithstanding their poverty, they are always chearful, and in high spirits.

August the 29th. BY the desire of the governor-general, marquis *de la Jonquiere*,

† It was probably impregnated with particles of copper ore.

and

and of marquis *de la Galiſſonniere*, I ſet out, with ſome *French* gentlemen, to viſit the pretended ſilver-mine, or the lead-mine, near the bay St. *Paul*. I was glad to undertake this journey, as it gave me an opportunity of ſeeing a much greater part of the country, than I ſhould otherwiſe have done. This morning therefore we ſet out on our tour in a boat, and went down the river St. *Lawrence*.

The harveſt was now at hand, and I ſaw all the people at work in the corn-fields. They had began to reap wheat and oats, a week ago.

The proſpect near *Quebec* is very lively from the river. The town lies very high, and all the churches, and other buildings, appear very conſpicuous. The ſhips in the river below ornament the landſcape on that ſide. The powder magazine, which ſtands at the ſummit of the mountain, on which the town is built, towers above all the other buildings.

The country we paſſed by afforded a no leſs charming ſight. The river St. *Lawrence* flows nearly from ſouth to north here; on both ſides of it are cultivated fields, but more on the weſt ſide than on the eaſt ſide. The hills on both ſhores are ſteep, and high. A number of fine hills,

hills, separated from each other, large fields, which looked quite white from the corn with which they are covered, and excellent woods of deciduous trees, made the country round us look very pleasant. Now and then we saw a church of stone, and in several places brooks fell from the hills into the river. Where the brooks are considerable, there they have made saw-mills, and water-mills.

AFTER rowing for the space of a *French* mile and a half, we came to the isle of *Orleans*, which is a large island, near seven *French* miles and a half long, and almost two of those miles broad, in the widest part. It lies in the middle of the river St. *Lawrence*, is very high, has steep and very woody shores. There are some places without trees, which have farm-houses below, quite close to the shore. The isle itself is well cultivated, and nothing but fine houses of stone, large corn-fields, meadows, pastures, woods of deciduous trees, and some churches built of stone, are to be seen on it.

WE went into that branch of the river which flows on the west side of the isle of *Orleans*, it being the shortest. It is reckoned about a quarter of a *French* mile broad, but ships cannot take this road, on account

account of the fand-banks, which ly here near the projecting points of land, and on account of the fhallownefs of the water, the rocks, and ftones at the bottom. The fhores on both fides ftill kept the fame appearance as before. On the weft fide, or on the continent, the hills near the river confift throughout of black lime-flate, and the houfes of the peafants are made of this kind of ftone, white-wafhed on the outfide. Some few houfes are of different kinds of ftone. The row of ten mountains, which is on the weft fide of the river, and runs nearly from fouth to north, gradually comes nearer to the river: for at *Quebec* they are near two *French* miles diftant from the fhore; but nine *French* miles lower down the river, they are almoft clofe to the fhore. Thefe mountains are generally covered with woods, but in fome places the woods have been deftroyed by accidental fires. About eight *French* miles and a half from *Quebec*, on the weft fide of the river, is a church, called St. *Anne*, clofe to the fhore. This church is remarkable, becaufe the fhips from *France* and other parts, as foon as they are got fo far up the river St. *Lawrence*, as to get fight of it, give a general difcharge of their artillery, as a fign of joy, that they have paft paſt all danger in the river, and have eſcaped all the ſands in it.

The water had a pale red colour, and was very dirty in thoſe parts of the river, which we ſaw to-day, though it was every where computed above ſix fathoms deep. Somewhat below St. *Anne,* on the weſt ſide of the river St. *Lawrence,* another river, called *la Grande Riviere,* or the *Great River,* falls in it. Its water flows with ſuch violence, as to make its way almoſt into the middle of the branch of the river St. *Lawrence,* which runs between the continent, and the iſle of *Orleans.*

About two o'clock in the afternoon the tide began to flow up the river, and the wind being likewiſe againſt us, we could not proceed any farther, till the tide began to ebb. We therefore took up our night lodgings in a great farm, belonging to the prieſts in *Quebec,* near which is a fine church, called St. *Joachim,* after a voyage of about eight *French* miles. We were exceeding well received here. The king has given all the country round about this place to the ſeminary, or the prieſts at *Quebec,* who have leaſed it to farmers, who have built houſes on it. Here are two prieſts, and a number of young boys, whom they inſtruct in reading, writing, and Latin.

tin. Moſt of theſe boys are deſigned for prieſts: Directly oppoſite this farm, to the eaſtward, is the north-eaſt point, or the extremity of the iſle of *Orleans*.

ALL the gardens in *Canada* abound with red currant ſhrubs, which were at firſt brought over from *Europe*. They grow exceſſively well here, and the ſhrubs, or buſhes, are quite red, being covered all over with the berries.

THE wild vines* grow pretty plentifully in the woods. In all other parts of *Canada* they plant them in the gardens, near arbours, and ſummer-houſes. The ſummer-houſes are made entirely of laths, over which the vines climb with their tendrils, and cover them entirely with their foliage, ſo as to ſhelter them entirely from the heat of the ſun. They are very refreſhing and cool, in ſummer.

THE ſtrong contrary winds obliged us to ly all night at St. *Joachim*.

Auguſt the 30th. THIS morning we continued our journey in ſpite of the wind, which was very violent againſt us. The water in the river begins to get a brackiſh taſte, when the tide is higheſt, ſomewhat below St. *Joachim*, and the further one

---

* Vitis labruſca & vulpina.

goes down, the more the saline taste encreases. At first the western shore of the river has fine, but low corn-fields, but soon after the high mountains run close to the river side. Before they come to the river the hilly shores consist of black lime-slate; but as soon as the high mountains appear on the river side, the lime-slates disappear. For the stone, of which the high mountains consist, is a chalky rock-stone, mixed with glimmer and quartz*. The glimmer is black; the quartz partly violet, and partly grey. All the four constituent parts are so well mixed together, as not to be easily separated by an instrument, though plainly distinguishable with the eye. During our journey to-day, the breadth of the river was generally three *French* miles. They shewed me the turnings the ships are obliged to sail in, which seem to be very troublesome, as they are obliged to bear away for either shore, as occasion requires, or as the rocks and sands in the river oblige them to do.

For the distance of five *French* miles we had a very dangerous passage to go through; for the whole western shore, along which we rowed, consists of very high and steep

---

* Saxum micaceo quarzoso-calcarium.

mountains, where we could not have found a single place to land with safety, during the space of five miles, in case a high wind had arisen. There are indeed two or three openings, or holes, in the mountains, into which one could have drawn the boat, in the greatest danger. But they are so narrow, that in case the boat could not find them in the hurry, it would inevitably be dashed against the rocks. These high mountains are either quite bare, or covered with some small firs, standing far asunder. In some places there are great clefts, going down the mountains, in which trees grow very close together, and are taller than on the other parts of the mountain; so that those places look like quick-hedges, planted on the solid rock. A little while after we passed a small church, and some farms round it. The place is called *Petite Riviere*, and they say, its inhabitants are very poor, which seems very probable. They have no more land to cultivate, than what lies between the mountains and the river, which in the widest part is not above three musket shot, and in most parts but one broad. About seventeen *French* miles from *Quebec* the water is so salty in the river, that no one can drink it, our rowers therefore provided themselves with a kettle full

full of fresh water this morning. About five o'clock in the evening, we arrived at bay St. *Paul*, and took our lodgings with the priests, who have a fine large house here, and entertained us very hospitably.

Bay St. *Paul* is a small parish, about eighteen *French* miles below *Quebec*, lying at some distance from the shore of a bay formed by the river, on a low plain. It is surrounded by high mountains on every side, one large gap excepted, which is over-against the river. All the farms are at some distance from each other. The church is reckoned one of the most ancient in *Canada*; which seems to be confirmed by its bad architecture, and want of ornaments; for the walls are formed of pieces of timber, erected at about two feet distance from each other, supporting the roof. Between these pieces of timber, they have made the walls of the church of lime-slate. The roof is flat. The church has no steeple, but a bell fixed above the roof, in the open air. Almost all the country in this neighbourhood belongs to the priests, who have leased it to the farmers. The inhabitants live chiefly upon agriculture and making of tar, which last is sold at *Quebec*.

This country being low, and situated upon a bay of the river, it may be conjectured,

tured, that this flat ground was formerly part of the bottom of the river, and formed itfelf, either by a decreafe of water in the river, or by an encreafe of earth, which was carried upon it from the continent by the brooks, or thrown on it by ftorms. A great part of the plants, which are to be met with here, are likewife marine; fuch as glafs-wort, fea milk-wort, and fea-fide peafe †. But when I have afked the inhabitants, whether they find fhells in the ground by digging for wells, they always anfwered in the negative. I received the fame anfwer from thofe who live in the low fields directly north of *Quebec*, and all agreed, that they never found any thing by digging, but different kinds of earth and fand.

It is remarkable, that there is generally a different wind in the bay from that in the river, which arifes from the high mountains, covered with tall woods, with which it is furrounded on every fide but one. For example, when the wind comes from the river, it ftrikes againft one of the mountains at the entrance of the bay, it is reflected, and confequently takes a direction quite different from what it had before.

† *Salicornia, Glaux, Pifum maritimum.*

I FOUND

I FOUND sand of three kinds upon the shore; one is a clear coarse sand, consisting of angulated grains of quartz, and is very common on the shore; the other is a fine black sand, which I have likewise found in abundance on the shores of lake *Champlain*,* and which is common all over *Canada*, Almost every grain of it is attracted by the magnet. Besides this, there is a granet coloured sand †, which is likewise very fine. This may owe its origin to the granet coloured grains of sand, which are to be found in all the stones and mountains here near the shore. The sand may have arisen from the crumbled pieces of some stones, or the stones may have been composed of it. I have found both this and the black sand on the shores, in several parts of this journey; but the black sand was always the most plentiful.

August the 31st. ALL the high hills in the neighbourhood sent up a smoke this morning, as from a charcoal-kiln.

GNATS are innumerable here; and as soon as one looks out of doors, they immediately attack him; and they are still worse in the woods. They are exactly the same

---

* See p. 24. of this volume.
† See p. 24. of this volume.

gnats as our common *Swedish* ones, being only somewhat less than the *North-American* gnats all are. Near Fort St. *Jean*, I have likewise seen gnats which were the same with ours, but they were somewhat bigger, almost of the size of our crane-flies*. Those which are here, are beyond measure blood-thirsty. However, I comforted myself, because the time of their disappearance was near at hand.

This afternoon we went still lower down the river St. *Lawrence*, to a place, where, we were told, there were silver or lead mines. Somewhat below bay St. *Paul*, we passed a neck of land, which consists entirely of a grey, pretty compact limestone, lying in dipping, and almost perpendicular strata. It seems to be merely a variety of the black lime-slates. The strata dip to the south-east, and basset out to the north-west. The thickness of each is from ten to fifteen inches. When the stone is broken, it has a strong smell, like stinkstone. We kept, as before, to the western shore of the river, which consists of nothing but steep mountains and rocks. The river is not above three *French* miles broad here. Now and then we could see stripes in the

* *Tipula hortorum.* Linn.

rock of a fine white, loose, semi-opaque spar. In some places of the river are pieces of rock as big as houses, which had rolled from the mountains in spring. The places they formerly occupied are plainly to be seen.

In several places, they have eel-traps in the river, like those I have before described †.

By way of amusement, I wrote down a few *Algonkin* words, which I learnt from a *Jesuit* who has been a long time among the *Algonkins*. They call water, *mukuman*; the head, *ustigon*; the heart, *uta*; the body, *veetras*; the foot, *ukhita*; a little boat, *ush*; a ship, *nabikoan*; fire, *skute*; hay, *maskoosee*; the hare, *whabus*; (they have a verb, which expresses the action of hunting hares, derived from the noun); the marten, *whabistanis*; the elk, *moosu* * (but so that the final *u* is hardly pronounced); the

---

† See p. 92. of this volume.

* The famous *moose-deer* is accordingly nothing but an elk; for no one can deny the derivation of *moose-deer* from *moosu*. Considering especially, that before the *Iroquese* or Five Nations grew to that power, which they at present have all over *North-America*, the *Algonkins* were then the leading nation among the *Indians*, and their language was of course then a most universal language over the greater part of *North-America*; and though they have been very nearly destroyed by the *Iroquese*, their language is still more universal in *Canada*, than any of the rest. F.

rein-

rein-deer, *atticku*; the moufe, *mawitulfis*. The *Jefuit* who told me thofe particulars, likewife informed me, that he had great reafon to believe, that, if any *Indians* here owed their origin to *Tataria*, he thought the *Algonkins* certainly did; for their language is univerfally fpoken in that part of *North-America*, which lies far to the weft of *Canada*, towards *Afia*. It is faid to be a very copious language; as for example, the verb *to go upon the ice*, is entirely different in the *Algonkin* from *to go upon dry land, to go upon the mountains*, &c.

LATE at night we arrived at *Terre d'Eboulement*, which is twenty-two *French* miles from *Quebec*, and the laft cultivated place on the weftern fhore of the river St. *Lawrence*. The country lower down is faid to be fo mountainous, that no body can live in it, there not being a fingle fpot of ground, which could be tilled. A little church, belonging to this place, ftands on the fhore, near the water.

No walnut-trees grow near this village, nor are there any kinds of them further north of this place. At bay St. *Paul*, there are two or three walnut-trees of that fpecies which the *Englifh* call butter-nut-trees; but they are looked upon as great rarities,

and

and there are no others in the neighbourhood.

Oaks of all kinds, will not grow near this place, nor lower down, or further north.

Wheat is the kind of corn which is sown in the greatest quantities here. The soil is prettty fertile, and they have sometimes got twenty-four or twenty-six bushels from one, though the harvest is generally ten or twelve fold. The bread here is whiter than any where else in *Canada*.

They sow plenty of oats, and it succeeds better than the wheat.

They sow likewise a great quantity of peas, which yield a greater encrease than any corn; and there are examples of its producing an hundred fold.

Here are but few birds; and those that pass the summer here, migrate in autumn; so that there are no other birds than snowbirds, red partridges, and ravens, in winter. Even crows do not venture to expose themselves to the rigours of winter, but take flight in autumn.

The *Bull-frogs* live in the pools of this neighbourhood. *Fire flies* are likewise to be found here.

Instead of candles, they make use of lamps in country places, in which they burn

burn train-oil of porpeſſes, which is the common oil here. Where they have none of it, they ſupply its place with train-oil of ſeals.

September the 1ſt. There was a woman with child in this village, who was now in the fifty-ninth year of her age. She had not had the catamenia during eighteen years. In the year 1748, ſhe got the ſmall-pox, and now ſhe was very big. She ſaid ſhe was very well, and could feel the motions of the fœtus. She looked very well, and had her huſband alive. This being an uncommon caſe, ſhe was brought to the royal phyſician, M. *Gaulthier*, who accompanied us on this journey.

At half an hour after ſeven this morning we went down the river. The country near *Terre d'Eboulement* is high, and conſiſts of hills of a looſe mould, which ly in three or four rows above each other, and are all well cultivated, and moſtly turned into corn-fields; though there are likewiſe meadows and paſtures.

The great earthquake which happened in *Canada*, in *February*, 1663, and which is mentioned by *Charlevoix* [*], has done conſiderable damage to this place. Many

---

[*] See his *Hiſtoire de la Nouvelle France*, Tom. II. p. m. 125.

hills tumbled down; and a great part of the corn-fields on the lowest hills were destroyed. They shewed me several little islands, which arose in the river on this occasion.

There are pieces of black lime-slate scattered on those hills, which consist of mould. For the space of eight *French* miles along the side of the river, there is not a piece of lime-slate to be seen; but instead of it, there are high grey mountains, consisting of a rock-stone, which contains a purple and a crystaline quartz, mixed with lime-stone, and black glimmer. The roots of these mountains go into the water. We now begin to see the lime-slates again.

Here are a number of Terns *, which fly about, and make a noise along the shore.

The river is here computed at about four *French* miles broad.

On the sides of the river, about two *French* miles inland, there are such terraces of earth as at *Terre d'Eboulement*; but soon after they are succeeded by high disagreeable mountains.

Several brooks fall into the river here, over the steep shores, with a great noise. The shores are sometimes several yards high,

* *Sterna hirundo* Linn.

high, and confift either of earth, or of rock-ftone.

One of thefe brooks, which flows over a hill of lime-ftone, contains a mineral water. It has a ftrong fmell of fulphur, is very clear, and does not change its colour, when mixed with gall-apples. If it is poured into a filver cup, it looks as if the cup was gilt; and the water leaves a fediment of a crimfon colour at the bottom. The ftones and pieces of wood, which ly in the water, are covered with a flime, which is pale grey at the top, and black at the bottom of the ftone. This flime has not much pungency, but taftes like oil of tobacco. My hands had a fulphureous fmell all day, becaufe I had handled fome of the flimy ftones.

The black lime-flate now abounds again, near the level of the water. It lies in ftrata, which are placed almoft perpendicularly near each other, inclining a little towards W. S. W. Each ftratum is between ten and fifteen inches thick. Moft of them are fhivered into thin leaves at the top, towards the day; but in the infide, whither neither fun, nor air and water can penetrate, they are clofe and compact. Some of thefe ftones are not quite black, but have a greyifh caft.

About noon we arrived at *Cap aux Oyes*, or *Geese Cape*, which has probably got its name from the number of wild geese which the *French* found near it, on their first arrival in *Canada*. At present, we saw neither geese, nor any kind of birds here, a single raven excepted. Here we were to examine the renowned metallic veins in the mountain; but found nothing more than small veins of a fine white spar, containing a few specks of lead ore. *Cap aux Oyes* is computed twenty-two, or twenty-five *French* miles distant from *Quebec*. I was most pleased by finding, that most of the plants are the same as grow in *Sweden*; a proof of which I shall produce in the sequel.

The sand-reed * grows in abundance in the sand, and prevents its being blown about by the wind.

The sea-lyme grass † likewise abounds on the shores. Both it and the preceding plant are called *Seigle de mer* ‡ by the *French*. I have been assured that these plants grow in great plenty in *Newfoundland*, and on other *North-American* shores; the places covered with them looking, at

---

* *Arundo arenaria* Linn.
† *Elymus arenarius* Linn.
‡ Sea-rye.

a disa distance, like corn-fields; which might explain the passage in our northern accounts, of the *excellent wine land* *, which mentions, that they had found whole fields of wheat growing wild.

THE sea-side plantain † is very frequent on the shore. The *French* boil its leaves in a broth on their sea-voyages, or eat them as a sallad. It may likewise be pickled like samphire.

THE bear-berries ‡ grow in great abundance here. The *Indians*, *French*, *English*, and *Dutch*, in those parts of *North-America*, which I have seen, call them *Sagackhomi*, and mix the leaves with tobacco for their use.

GALE, or sweet willow §, is likewise abundant here. The *French* call it *Laurier*, and some *Poivrier*. They put the leaves into their broth, to give it a pleasant taste.

THE sea-rocket ‖ is, likewise, not un-

---

* *Vinland det grda*, or the good wine-land, is the name which the old *Scandinavian* navigators gave to *America*, which they discovered long before *Columbus*. See *Torfæi Historia Vinlandiæ antiquæ f. partis Americæ septentrionalis*. Hafniæ 1715, 4to. and Mr. *George Westmann's*, A. M. Dissertation on that Subject. Abo 1747. F.

† *Plantago maritima* Linn.
‡ *Arbutus uva ursi* Linn.
§ *Myrica gale* Linn.
‖ *Bunias cakile* Linn.

common. Its root is pounded, mixed with flour, and eaten here, when there is a scarcity of bread.

THE sorb-tree, or mountain-ash, the cranberry-bush, the juniper-tree, the sea-side pease, the *Linnæa*, and many other *Swedish* plants, are likewise to be met with here.

WE returned to bay *St. Paul* to-day. A grey seal swam behind the boat for some time, but was not near enough to be shot at.

September the 2d. THIS morning we went to see the silver or lead veins. They ly a little on the south-side of the mills, belonging to the priests. The mountain in which the veins ly, has the same constituent parts, as the other high grey rocks in this place, viz. a rock-stone composed of a whitish or pale grey lime-stone, a purple or almost garnet-coloured quartz, and a black glimmer. The lime-stone is in greater quantities here than the other parts; and it is so fine as to be hardly visible. It effervesces very strongly with *aqua fortis*. The purple or garnet-coloured quartz is next in quantity; lies scattered in exceeding small grains, and strikes fire when struck with a steel. The little black particles of glimmer follow next; and last of all, the transparent crystalline speckles of quartz.

quartz. There are some small grains of spar in the lime-stone. All the different kinds of stone are very well mixed together, except that the glimmer now and then forms little veins and lines. The stone is very hard; but when exposed to sun-shine and the open air, it changes so much as to look quite rotten, and becomes friable; and in that case, its constituent particles grow quite undistinguishable. The mountain is quite full of perpendicular clefts, in which the veins of lead-ore run from E. S. E. to W. N. W. It seems the mountain had formerly got cracks here, which were afterwards filled up with a kind of stone, in which the lead-ore was generated. That stone which contains the lead-ore is a soft, white, often semidiaphanous spar, which works very easily. In it there are sometimes stripes of a snowy white lime-stone, and almost always veins of a green kind of stone like quartz. This spar has many cracks, and divides into such pieces as quartz; but is much softer, never strikes fire with steel, does not effervesce with acids, and is not smooth to the touch. It seems to be a species of Mr. Professor *Wallerius's* vitrescent spar *.

* See *Wallerius's* Mineralogy, *Germ.* ed. p. 87, F.,?. Introd. to Mineralogy, p. 13.

There are sometimes small pieces of a greyish quartz in this spar, which emit strong sparks of fire, when struck with a steel. In these kinds of stone the lead ore is lodged. It commonly lies in little lumps of the size of peas; but sometimes in specks of an inch square, or bigger. The ore is very clear, and lies in little cubes *. It is generally very poor, a few places excepted. The veins of soft spar, and other kinds of stone, are very narrow, and commonly from ten to fifteen inches broad. In a few places they are twenty inches broad; and in one single place twenty-two and a half. The brook which intersects the mountain towards the mills, runs down so deep into the mountain, that the distance from the summit of the hill, to the bottom of the brook, is near twelve yards. Here I examined the veins, and found that they always keep the same breadth, not encreasing near the bottom of the brook; and likewise, that they are no richer below, than at the top. From hence it may be easily concluded, that it is not worth while sinking mines here. Of these veins there are three or four in this neighbourhood, at some distance from each other,

---

* It is a *cubic lead ore*, or *lead glance*. *Forster's* Introd. to Mineralogy, p. 51.

but all of the fame quality. The veins are almoſt perpendicular, ſometimes deviating a little. When pieces of the green ſtone before mentioned ly in the water, a great deal of the adherent white ſpar and lime-ſtone is confumed; but the green ſtone remains untouched. That part of the veins which is turned towards the air is always very rough, becauſe the fun, air, and rain, have mouldered a great part of the ſpar and lime-ſtone; but the green ſtone has reſiſted their attacks. They ſometimes find deep holes in theſe veins, filled with mountain cryſtals. The greateſt quantity of lead or ſilver ore is to be found next to the rock, or even on the ſides of the vein. There are now and then little grains of pyrites in the ſpar, which have a fine gold colour. The green ſtone when pounded, and put on a red-hot ſhovel, burns with a blue flame. Some fay, they can then obſerve a ſulphureous ſmell, which I could never perceive, though my ſenſe of ſmelling is very perfect. When this green ſtone is grown quite red-hot, it loſes its green colour, and acquires a whitiſh one, but will not effervefce with *aqua fortis*.

THE ſulphureous ſprings (if I may fo call them) are at the foot of the mountain, which contains the ſilver, or lead ore. Several

veral springs join here, and form a little brook. The water in those brooks is covered with a white membrane, and leaves a white, mealy matter on the trees, and other bodies in its way; this matter has a strong sulphureous smell. Trees, covered with this mealy matter, when dried and set on fire, burn with a blue flame, and emit a smell of sulphur. The water does not change by being mixed with gall-apples, nor does it change blue paper into a different colour, which is put into it. It makes no good lather with soap. Silver is tarnished, and turns black, if kept in this water for a little while. The blade of a knife was turned quite black, after it had lain about three hours in it. It has a disagreeable smell, which, they say, it spreads still more in rainy weather. A number of grashoppers were fallen into it at present. The inhabitants used this water, as a remedy against the itch.

IN the afternoon we went to see another vein, which had been spoken of as silver ore. It lies about a quarter of a mile to the north-east of bay St. *Paul*, near a point of land called *Cap au Corbeau*, close to the shore of the river St. *Lawrence*. The mountain in which these veins ly, consist of a pale red vitrescent spar, a black glimmer,

mer, a pale lime-ſtone, purple or garnet-coloured grains of quartz, and ſome tranſparent quartz. Sometimes the reddiſh vitreſcent ſpar is the moſt abundant, and lies in long ſtripes of ſmall hard grains. Sometimes the fine black glimmer abounds more than the remaining conſtituent parts; and theſe two laſt kinds of ſtone generally run in alternate ſtripes. The white lime-ſtone which conſiſts of almoſt inviſible particles, is mixed in among them. The garnet-coloured quartz grains appear here and there, and ſometimes form whole ſtripes. They are as big as pin's heads, round, ſhining, and ſtrike fire with ſteel. All theſe ſtones are very hard, and the mountains near the ſea, conſiſt entirely of them. They ſometimes ly in almoſt perpendicular ſtrata, of ten or fifteen inches thickneſs. The ſtrata, however, point with their upper ends to the north-weſt, and go upwards from the river, as if the water, which is cloſe to the ſouth-eaſt ſide of the mountains, had forced the ſtrata to lean on that ſide. Theſe mountains contain very narrow veins of a white, and ſometimes of a greeniſh, fine, ſemidiaphanous, ſoft ſpar, which crumbles eaſily into grains. In this ſpar they very frequently find ſpecks, which look like a calamine blend.

blend *. Now and then, and but very seldom, there is a grain of lead-ore. The mountains near the shore consist sometimes of a black fine-grained horn-stone, and a ferruginous lime-stone. The horn-stone in that case is always in three or four times as great a quantity as the lime-stone.

In this neighbourhood there is likewise a sulphureous spring, having exactly the same qualities as that which I have before described. The broad-leaved *Reed Mace* † grows in the very spring, and succeeds extremely well. A mountain-ash stood near it, whose berries were of a pale yellow fading colour, whereas on all other mountain-ashes they have a deep red colour.

They make great quantities of tar at bay St. *Paul*. We now passed near a place in which they burn tar, during summer. It is exactly the same with ours in *East-Bothnia*, only somewhat less; though I have been told, that there are sometimes very great manufactures of it here. The tar is made solely of the *Pin rouge* ‡, or red Pine. All other firs, of which here are several kinds, are not fit for this purpose,

---
* *Forster's* Introd. to Mineralogy. p. 50. *Zincum sterilum*, Linn. Syst. Nat. III. p. 126. Ed. XII.

† *Typha latifolia*, Linn.

‡ Pinus foliis geminis longis; ramis triplici fasciculo foliorum terminatis, conis ovatis lævibus. *Flor. Canad.*

pose, because they do not give tar enough to repay the trouble the people are at. They make use of the roots alone, which are quite full of resin, and which they dig out of the ground; and of about two yards of the stem, just above the root, laying aside all the rest. They have not yet learnt the art of drawing the resin to one side of the tree, by peeling off the bark; at least they never take this method. The tarbarrels are but about half the size of ours. A ton holds forty-six pots, and sells at present for twenty-five *francs* at *Quebec*. The tar is reckoned pretty good.

The sand on the shore of the river St. *Lawrence*, consists in some places of a kind of pearl-sand. The grains are of quartz, small and semidiaphanous. In some places it consists of little particles of glimmer; and there are likewise spots, covered with the garnet-coloured sand, which I have before described, and which abounds in *Canada*.

September the 4th. The mountains hereabouts were covered with a very thick fog to-day, resembling the smoak of a charcoal kiln. Many of these mountains are very high. During my stay in *Canada*, I asked many people, who have travelled much in *North-America*, whether they ever met with mountains so high, that the snow never melts on them in winter; to which
they

they always answered in the negative. They say that the snow sometimes stays on the highest, *viz.* on some of those between *Canada* and the *English* colonies, during a great part of the summer; but that it melts as soon as the great heat begins.

EVERY countryman sows as much flax as he wants for his own use. They had already taken it up some time ago, and spread it on the fields, meadows, and pastures, in order to bleach it. It was very short this year in *Canada*.

THEY find iron-ore in several places hereabouts. Almost a *Swedish* mile from bay St. *Paul*, up in the country, there is a whole mountain full of iron ore. The country round it is covered with a thick forest, and has many rivulets of different sizes, which seem to make the erection of iron-works very easy here. But the government having as yet suffered very much by the iron-works at *Trois Rivieres*, nobody ventures to propose any thing further in that way.

September the 5th. EARLY this morning we set out on our return to *Quebec*. We continued our journey at noon, notwithstanding the heavy rain and thunder we got afterwards. At that time we were just juft at *Petite Riviere*, and the tide beginning to ebb, it was impoffible for us to come up againft it; therefore we lay by here, and went on fhore.

Petite Riviere is a little village, on the weftern fide of the river St. *Lawrence*, and lies on a little rivulet, from whence it takes its name. The houfes are built of ftone, and are difperfed over the country. Here is likewife a fine little church of ftone. To the weft of the village are fome very high mountains, which caufe the fun to fet three or four hours fooner here, than ordinary. The river St. *Lawrence* annually cuts off a piece of land, on the eaft fide of the village, fo that the inhabitants fear they will in a fhort time lofe all the land they poffefs here, which at moft is but a mufket fhot broad. All the houfes here are very full of children.

THE lime-flates on the hills are of two kinds. One is a black one, which I have often mentioned, and on which the town of *Quebec* is built. The other is generally black, and fometimes dark grey, and feems to be a fpecies of the former. It is called *Pierre à chaux* here. It is chiefly diftinguifhed from the former, by being cut very eafily, giving a very white lime, when burnt, and not eafily mouldering into fhivers

vers in the air. The walls of the houses here are entirely made of this slate; and likewise the chimnies, those places excepted, which are exposed to the greatest fire, where they place pieces of grey rock-stone, mixed with a deal of glimmer. The mountains near *Petite Riviere* consist merely of a grey rock-stone, which is entirely the same with that which I described near the lead-mines of bay St. *Paul*. The foot of these mountains consists of one of the lime-slate kinds. A great part of the *Canada* mountains of grey rock-stone stand on a kind of slate, in the same manner as the grey rocks of *West-Gothland* in *Sweden*.

September the 6th. THEY catch eels and porpesses here, at a certain season of the year, *viz*. at the end of *September*, and during the whole month of *October*. The eels come up the river at that time, and are caught in the manner I have before described. They are followed by the porpesses, which feed upon them. The greater the quantity of eels is, the greater is likewise the number of porpesses, which are caught in the following manner. When the tide ebbs in the river, the porpesses commonly go down along the sides of the river, catching the eels which they find there.

there. The inhabitants of this place therefore stick little twigs, or branches with leaves, into the river, in a curve line or arch, the ends of which look towards the shore, but stand at some distance from it, leaving a passage there. The branches stand about two feet distant from each other. When the porpesses come amongst them, and perceive the rustling the water makes with the leaves, they dare not venture to proceed, fearing lest there should be a snare, or trap, and endeavour to go back. Mean while the water has receded so much, that in going back they light upon one of the ends of the arch, whose moving leaves frighten them again. In this confusion they swim backwards and forwards, till the water is entirely ebbed off, and they ly on the bottom, where the inhabitants kill them. They give a great quantity of train-oil.

NEAR the shore, is a grey clay, full of ferruginous cracks, and pierced by worms. The holes are small, perpendicular, and big enough to admit a middling pin. Their sides are likewise ferruginous, and half-petrified; and where the clay has been washed away by the water, the rest looks like ocker-coloured stumps of tobacco-pipe tubes.

AT

At noon we left *Petite Riviere*, and continued our journey towards St. *Joachim*.

Between *Petite Riviere*, which lies in a little bay, and St. *Joachim*, the western shore of the river St. *Lawrence* consists of prominent mountains, between which there are several small bays. They have found, by long experience, that there is always a wind on these mountains, even when it is calm at *Petite Riviere*. And when the wind is pretty high at the last-mentioned place, it is not adviseable to go to *Quebec* in a boat, the wind and waves, in that case, being very high near these mountains. We had at present an opportunity of experiencing it. In the creeks between the mountains, the water was almost quite smooth; but on our coming near one of the points formed by the high mountains, the waves encreased, and the wind was so high, that two people were forced to take care of the helm, and the mast broke several times. The waves are likewise greatly encreased by the strong current near those points or capes.

September the 7th. A little before noon, we continued our voyage from St. *Joachim*.

They employ tree-mushrooms very frequently instead of tinder. Those which are

are taken from the sugar-maple are reckoned the best; those of the red maple are next in goodness; and next to them, those of the sugar-birch. For want of these, they likewise make use of those which grow on the asp-tree or tremble.

THERE are no other ever-green trees in this part of *Canada*, than the thuya, the yew, and some of the fir kind.

THE thuya is esteemed for resisting putrefaction much longer than any other wood; and next in goodness to it is the pine, called *perusse* here.

THEY make cheese in several places hereabouts. That of the isle of *Orleans* is, however, reckoned the best. This kind is small, thin, and round; and four of them weigh about a *French* pound. Twelve of them sell for thirty sols. A pound of salt butter costs ten sols at *Quebec*, and of fresh butter, fifteen sols. Formerly, they could get a pound of butter for four sols here.

THE corn-fields towards the river are sloping; they are suffered to ly fallow and to be sown alternately. The sown ones looked yellow at this distance, and the fallow ones green. The weeds are left on the latter all summer, for the cattle to feed upon.

THE ash wood furnishes the best hoops for tuns here; and for want of it, they take the thuya, little birch-trees, wild cherry-trees, and others.

The hills near the river, on the western side, opposite the isle of *Orleans*, are very high and pretty steep. They consist, in most part, of black lime-slate. There are likewise some spots which consist of a rock-stone, which, at first sight, looks like a sand-stone, and is composed of grey quartz, a reddish lime-stone, a little grey lime-stone, and some pale grey grains of sand. These parts of the stone are small and pretty equally mixed with each other. The stone looks red, with a greyish cast, and is very hard. It lies in strata, one above another. The thickness of each stratum is about five inches. It is remarkable, that there are both elevated and hollow impressions of pectinites on the surface, where one likewise meets with the petrified shells themselves; but on breaking the stone, it does not even contain the least vestige of an impression or petrified shell. All the impressions are small, about the length and breadth of an inch. The particles of quartz in the stone strike fire with steel, and the particles of lime-stone effervesce strongly with *aqua-fortis*. The upper and lower surfaces of the strata consist of lime-stone, and the inner parts of quartz. They break great quantities of this stone in order to build houses of it, pave floors with it, and make stair-cases of it. Great quantities of it are sent to *Quebec*. It is remarkable, that there are petrefactions in this stone, but never any in the black lime-slates

The women dye their woollen yarn yellow with seeds of gale,* which is called *poivrier* here, and grows abundant in wet places.

This evening, M. *Gaulthier* and I went to see the water-fall at *Montmorenci*. The country near the river is high and level, and laid out into meadows. Above them the high and steep hills begin, which are covered with a crust of mould, and turned into corn-fields. In some very steep places, and near the rivulets, the hills consist of mere black lime-slate, which is often crumbled into small pieces, like earth. All the fields below the hills are full of such pieces of lime-slate. When some of the larger pieces are broken, they smell like stink-stone. In some more elevated places, the earth consists of a pale red colour; and the lime-slates are likewise reddish.

The water-fall near *Montmorenci* is one of the highest I ever saw. It is in a river

* *Myrica gale.* Linn.

whose breadth is not very considerable, and
falls over the steep side of a hill, consisting
entirely of black lime-slate. The fall is
now at the bottom of a little creek of the
river. Both sides of the creek consist merely of black lime-slate, which is very much
cracked and tumbled down. The hill of
lime-slate under the water-fall is quite perpendicular, and one cannot look at it without astonishment. The rain of the preceding days had encreased the water in the
river, which gave the fall a grander appearance. The breadth of the fall is not above
ten or twelve yards. Its perpendicular height
Mr. *Gaulthier* and I guessed to be between
a hundred and ten and a hundred and twenty feet; and on our return to *Quebec*, we
found our guess confirmed by several gentlemen, who had actually measured the fall,
and found it to be nearly as we had conjectured. The people who live in the neighbourhood exaggerate in their accounts of it,
absolutely declaring that it is three hundred
feet high. Father *Charlevoix*† is too sparing
in giving it only forty feet in height. At
the bottom of the fall, there is always a
thick fog of vapours, spreading about the
water, being resolved into them by its vio-

---

† See his *Histoire de la Nouv. France*, tom. v. p. m. 100.

lent

lent fall. This fog occafions almoft perpetual rain here, which is more or lefs heavy, in proportion to its diftance from the fall. Mr. *Gaulthier* and myfelf, together with the man who fhewed us the way, were willing to come nearer to the falling water, in order to examine more accurately how it came down from fuch a height, and how the ftone behind the water looked. But, being about twelve yards off the fall, a fudden guft of wind blew a thick fog upon us, which, in lefs than a minute, had wet us as thoroughly as if we had walked for half an hour in a heavy fhower. We therefore hurried away as faft as we could, and were glad to get off. The noife of the fall is fometimes heard at *Quebec*, which is two *French* miles off to the fouthward; and this is a fign of a north-eaft wind. At other times, it can be well heard in the villages, a good way lower to the north; and it is then reckoned an undoubted fign of a fouth-weft wind, or of rain. The black lime-flate on the fides of the fall lies in dipping, and almoft perpendicular ftrata. In thefe lime-flate ftrata, are the following kinds of ftone to be met with.

*Fibrous gypfum.** This lies in very thin

---

* *Gypfum amiantiforme, Waller. Min. Germ.* ed. p. 74. *Fibrous or radiated gypfum, Forft. Introd. to Mineralogy,* p. 16,

leaves

leaves between the cracks of the lime-flate. Its colour is a fnowy white. I have found it in feveral parts of *Canada,* in the fame black lime-ftone.

Pierre à Calumet. This is the *French* name of a ftone difpofed in ftrata between the lime-flate, and of which they make almoft all the tobacco-pipe heads in the country. The thicknefs of the ftrata is different. I have feen pieces near fifteen inches thick; but they are commonly between four and five inches thick. When the ftone is long expofed to the open air or heat of the fun, it gets a yellow colour; but in the infide it is grey. It is a lime-ftone of fuch a compactnefs, that its particles are not diftinguifhable by the naked eye. It is pretty foft, and will bear cutting with a knife. From this quality, the people likewife judge of the goodnefs of the ftone for tobacco-pipe heads; for the hard pieces of it are not fo fit for ufe as the fofter ones. I have feen fome of thefe ftones fhivering into thin leaves on the outfide where they were expofed to the fun. All the tobacco-pipe heads, which the common people in *Canada* make ufe of, are made of this ftone, and are ornamented in different ways. A great part of the gentry likewife make ufe of them, efpecially when they are on a journey.

The

The *Indians* have employed this stone for the same purposes for several ages past, and have taught it the *Europeans*. The heads of the tobacco-pipes are naturally of a pale grey colour; but they are blackened whilst they are quite new, to make them look better. They cover the head all over with grease, and hold it over a burning candle, or any other fire, by which means it gets a good black colour, which is encreased by frequent use. The tubes of the pipes are always made of wood†.

There are no coals near this fall, or in the steep hills close to it. However, the people in the neighbouring village shewed me a piece of coal, which, they said, they had found on one of the hills about the fall.

We arrived at *Quebec* very late at night.

September the 8th. Intermitting fevers of all kinds are very rare at *Quebec*, as Mr. *Gaulthier* affirms. On the contrary,

---

† All over *Poland*, *Russia*, *Turky*, and *Tartary*, they smoke out of pipes made of a kind of stone marle, to which they fix long wooden tubes; for which latter purpose, they commonly employ the young shoots of the various kinds of *spiræa*, which have a kind of pith easily to be thrust out. The stone-marle is called generally se:-lcum, being pretty soft; and by the *Tartars*, in C im a, it is called k ff kil. And as it cuts so easily, various figures are curiously carved in it, when it is worked into pipe-heads, which often are mounted with silver. F.

they are very common near Fort St. *Frederic*, and near Fort *Detroit*, which is a *French* colony, between lake *Erie* and lake *Huron*, in forty-three degrees north latitude.

Some of the people of quality make use of ice-cellars, to keep beer cool in, during summer, and to keep fresh flesh, which would not keep long in the great heat. These ice-cellars are commonly built of stone, under the house. The walls of it are covered with boards, because the ice is more easily consumed by stones. In winter, they fill it with snow, which is beat down with the feet, and covered with water. They then open the cellar holes and the door, to admit the cold. It is customary in summer to put a piece of ice into the water or wine which is to be drank.

All the salt which is made use of here, is imported from *France*. They likewise make good salt here of the sea water; but *France* keeping the salt trade entirely to itself, they do not go on with it here.

The *Esquimaux* are a particular kind of *American* savages, who live only near the water, and never far in the country, on *Terra Labrador*, between the most outward point of the mouth of the river St. *Lawrence* and *Hudson*'s bay. I have never had

an

an opportunity of seeing one of them. I have spoken with many *Frenchmen* who have seen them, and had them on board their own vessels. I shall here give a brief history of them, according to their unanimous accounts.

The *Esquimaux* are entirely different from the *Indians* of *North-America*, in regard to their complexion and their language. They are almost as white as *Europeans*, and have little eyes: the men have likewise beards. The *Indians*, on the contrary, are copper-coloured, and the men have no beards. The *Esquimaux* language is said to contain some *European* words.† Their houses are either caverns or clefts in the mountains, or huts of turf above ground. They never sow or plant vegetables, living chiefly on various kinds of whales, on seals,* and walrusses‡. Sometimes they likewise

---

† The *Moravian* brethren in *Greenland*, coming once over with some *Greenlanders* to *Terra Labrador*, the *Esquimaux* ran away at their appearance; but they ordered one of their *Greenlanders* to call them back in his language. The *Esquimaux* hearing his voice, and understanding the language, immediately stopped, came back, and were glad to find a countryman, and wherever they went, among the other *Esquimaux*, they gave out, that one of their brethren was returned. This proves the *Esquimaux* to be of a tribe different from any *European* nation, as the *Greenland* language has no similarity with any language in *Europe*. F.

* *Phoca vitulina*. Linn.
‡ *Trichechus rosmarus*. Linn.

catch

catch land animals, on which they feed. They eat moſt of their meat quite raw. Their drink is water; and people have likewiſe ſeen them drinking the ſea water, which was like brine.

Their ſhoes, ſtockings, breeches, and jackets are made of ſeal-ſkins well prepared, and ſewed together with nerves of whales, which may be twiſted like threads and are very tough. Their cloaths, the hairy ſide of which is turned outwards, are ſewed together ſo well, that they can go up to their ſhoulders in the water without wetting their under cloaths. Under their upper cloaths, they wear ſhirts and waiſtcoats made of ſeals ſkins, prepared ſo well as to be quite ſoft. I ſaw one of their womens dreſſes; a cap, a waiſtcoat, and coat, made all of one piece of ſeals ſkin well prepared, ſoft to the touch, and the hair on the outſide. Their is a long train behind at their coats, which ſcarce reach them to the middle of the thigh before; under it they wear breeches and boots, all of one piece. The ſhirt I ſaw was likewiſe made of a very ſoft ſeals ſkin. The *Eſquimaux* women are ſaid to be handſomer than any of the *American Indian* women, and their huſbands are accordingly more jealous in proportion.

I have

I HAVE likewife feen an *Efquimaux* boat. The outfide of it confifts entirely of fkins, the hair of which has been taken off; and the fides of the fkins on which they were, inferted are turned outwards, and feel as fmooth as vellum. The boat was near fourteen feet long, but very narrow, and very fharp pointed at the extremities. In the infide of the boat, they place two or three thin boards, which give a kind of form to the boat. It is quite covered with fkins at the top, excepting, near one end, a hole big enough for a fingle perfon to fit and row in, and keep his thighs and legs under the deck. The figure of the hole refembles a femi-circle, the bafe or diameter of which is turned towards the larger end of the boat. The hole is furrounded with wood, on which a foft folded fkin is faftened, with ftraps at its upper end. When the *Efquimaux* makes ufe of his boat, he puts his legs and thighs under the deck, fits down at the bottom of the boat, draws the fkin before mentioned round his body, and faftens it well with the ftraps; the waves may then beat over his boat with confiderable violence, and not a fingle drop comes into it; the cloaths of the *Efquimaux* keep the wet from him. He has an oar in his hand, which has a paddle at each end; it ferves him for

rowing

rowing with, and keeping the boat in equilibrium during a ftorm. The paddles of the oar are very narrow. The boat will contain but a fingle perfon. *Efquimaux* have often been found fafe in their boats many miles from land, in violent ftorms, where fhips found it difficult to fave themfelves. Their boats float on the waves like bladders, and they row them with incredible velocity. I am told, they have boats of different fhapes. They have likewife larger boats of wood, covered with leather in which feveral people may fit, and in which their women commonly go to fea.

Bows and arrows, javelins and harpoons, are their arms. With the laft they kill whales, and other large marine animals. The points of their arrows and harpoons are fometimes made of iron, fometimes of bone, and fometimes of the teeth of the walrufs. Their quivers are made of feals fkins. The needles with which they fow their cloaths are likewife made of iron or of bone. All their iron they get by fome means or other from the *Europeans*.

THEY fometimes go on board the *European* fhips in order to exchange fome of their goods for knives and other iron. But it is not advifeable for *Europeans* to go on fhore, unlefs they be numerous; for the

Ef-

Esquimaux are falſe and treacherous, and cannot ſuffer ſtrangers amongſt them. If they find themſelves too weak, they run away at the approach of ſtrangers; but if they think they are an over-match for them, they kill all that come in their way, without leaving a ſingle one alive. The *Europeans*, therefore, do not venture to let a greater number of *Esquimaux* come on board their ſhips than they can eaſily maſter. If they are ſhip-wrecked on the *Esquimaux* coaſts, they may as well be drowned in the ſea as come ſafe to the ſhore: this many *Europeans* have experienced. The *European* boats and ſhips which the *Esquimaux* get into their power, are immediately cut in pieces and robbed of all their nails and other iron, which they work into knives, needles, arrow-heads, &c. They make uſe of fire for no other purpoſes but working of iron, and preparing the ſkins of animals. Their meat is eaten all raw. When they come on board an *European* ſhip, and are offered ſome of the ſailors meat, they never will taſte of it till they have ſeen ſome *Europeans* eat it. Though nothing pleaſed other ſavage nations ſo much as brandy, yet many *Frenchmen* have aſſured me, that they never could prevail on the *Esquimaux* to take a dram of it. Their miſtruſt of other nations

is

the cause of it; for they undoubtedly imagine, that they are going to poison them, or do them some hurt; and I am not certain, whether they do not judge right. They have no ear-rings, and do not paint the face like the *American Indians*. For many centuries past, they have had dogs, whose ears are erected, and never hang down. They make use of them for hunting, and instead of horses in winter, for drawing their goods on the ice. They themselves sometimes ride in sledges drawn by dogs. They have no other domestic animal. There are, indeed, plenty of reindeer in their country; but it is not known, that either the *Esquimaux*, or any of the *Indians* in *America*, have ever tamed them. The *French* in *Canada*, who are in a manner the neighbours of the *Esquimaux*, have taken a deal of pains to carry on some kind of trade with them, and to endeavour to engage them to a more friendly intercourse with other nations. For that purpose, they took some *Esquimaux* children, taught them to read, and educated them in the best manner possible. The intention of the *French* was, to send these children to the *Esquimaux* again, that they might inform them of the kind treatment the *French* had given them, and thereby incline them to

con-

conceive a better opinion of the *French*. But unhappily all the children died of the small-pox, and the scheme was dropt. Many persons in *Canada* doubted, whether the scheme would have succeeded, though the children had been kept alive. For they say, there was formerly an *Esquimaux* taken by the *French*, and brought to *Canada*, where he staid a good while, and was treated with great civility. He learnt *French* pretty well, and seemed to relish the *French* way of living very well. When he was sent back to his countrymen, he was not able to make the least impression on them, in favour of the *French*; but was killed by his nearest relations, as half a *Frenchman* and foreigner. This inhuman proceeding of the *Esquimaux* against all strangers, is the reason why none of the *Indians* of *North America* ever give quarter to the *Esquimaux* if they meet with them, but kill them on the spot; though they frequently pardon their other enemies, and incorporate the prisoners into their nation.

For the use of those, who are fond of comparing the languages of several nations, I have here inserted a few *Esquimaux* words, communicated to me by the Jesuit *Saint Pie*. One, *kombuc*; two, *tigal*; three, *ké*; four, *miffilagat*; water, *fillalokto*; rain, *killa-*

killaluck; heaven, *taktuck*, or *nabugakſhe*; the ſun, *ſhikonak*, or *ſakaknuk*; the moon, *takock*; an egg, *manneguk*; the boat, *kagack*; the oar, *pacotick*; the knife, *ſhavié*; a dog, *mekké*, or *timilok*; the bow, *petikſick*; an arrow, *katſo*; the head, *niakock*; the ear, *tchiu*; the eye, *killik*, or *ſhik*; the hair, *nutſhad*; a tooth, *ukak*; the foot, *itikat*. Some think that they are nearly the ſame nation with the *Greenlanders*, or *Skralingers*; and pretend that there is a great affinity in the language *.

PLUMB-TREES of different ſorts, brought over from *France*, ſucceed very well here. The preſent year they did not begin to flower till this month. Some of them looked very well; and I am told the winter does not hurt them.

September the 11th. THE marquis *de la Galiſſonniere* is one of the three noblemen, who, above all others, have gained high eſteem with the *French* admiralty in the laſt war. They are the marquiſſes *de la Galiſſonniere, de la Jonquiere*, and *de l'E-*

---

* The above account of the *Eſquimanx* may be compared with *Henry Ellis's Account of a Voyage to Hudſon's Bay, by the Dobbs Galey and California*, &c. and *The Account of a Voyage for the Diſcovery of a North Weſt Paſſage by Hudſon's Streights, by the Clerk of the California*. Two Vols. 8vo. And laſtly, with *Crantz's Hiſtory of Greenland*. Two Vols. 8vo. F.

tendue.

tendure. The first of these was now above fifty years of age, of a low stature, and somewhat hump-backed, but of a very agreeable look. He had been here for some time as governor-general; and was going back to *France* one day this month. I have already mentioned something concerning this nobleman; but when I think of his many great qualities, I can never give him a sufficient encomium. He has a surprizing knowledge in all branches of science, and especially in natural history; in which he is so well versed, that when he began to speak with me about it, I imagined I saw our great *Linnæus* under a new form. When he spoke of the use of natural history, of the method of learning, and employing it to raise the state of a country, I was astonished to see him take his reasons from politics, as well as natural philosophy, mathematics, and other sciences. I own, that my conversation with this nobleman was very instructive to me; and I always drew a deal of useful knowledge from it. He told me several ways of employing natural history to the purposes of politics, and to make a country powerful, in order to depress its envious neighbours. Never has natural history had a greater promoter in this country; and it is very doubtful whether it will ever have his equal here. As foon as he got the place of governor-general, he began to take thofe meafures for getting information in natural hiftory, which I have mentioned before. When he faw people, who had for fome time been in a fettled place of the country, efpecially in the more remote parts, or had travelled in thofe parts, he always queftioned them about the trees, plants, earths, ftones, ores, animals, &c. of the place. He likewife enquired what ufe the inhabitants made of thefe things; in what ftate their hufbandry was ; what lakes, rivers, and paffages there are ; and a number of other particulars. Thofe who feemed to have clearer notions than the reft, were obliged to give him circumftantial defcriptions of what they had feen. He himfelf wrote down all the accounts he received ; and by this great application, fo uncommon among perfons of his rank, he foon acquired a knowledge of the moft diftant parts of *America*. The priefts, commandants of forts, and of feveral diftant places, are often furprized by his queftions, and wonder at his knowledge, when they come to *Quebec* to pay their vifits to him ; for he often tells them that near fuch a mountain, or on fuch a fhore, &c. where they often went a

hunting,

hunting, there are some particular plants, trees, earths, ores, &c. for he had got a knowledge of those things before. From hence it happened, that some of the inhabitants believed he had a preternatural knowledge of things, as he was able to mention all the curiosities of places, sometimes near two hundred *Swedish* miles from *Quebec*, though he never was there himself. Never was there a better statesman than he; and nobody can take better measures, and choose more proper means for improving a country, and encreasing its welfare. *Canada* was hardly acquainted with the treasure it possessed in the person of this nobleman, when it lost him again; the king wanted his services at home, and could not leave him so far off. He was going to *France* with a collection of natural curiosities; and a quantity of young trees and plants, in boxes full of earth.

THE black lime-slate has been repeatedly mentioned during the course of my journey. I will here give a more minute detail of it. The mountain on which *Quebec* is built, and the hills along the river St. *Lawrence*, consist of it for some miles together, on both sides of *Quebec*. About a yard from the surface, this stone is quite compact, and without any cracks; so that one cannot perceive that it is a flate, its particles being imperceptible. It lies in ſtrata, which vary from three or four inches, to twenty thick, and upwards. In the mountains on which *Quebec* is built, the ſtrata do not ly horizontal, but dipping, ſo as to be nearly perpendicular; the upper ends pointing north-weſt, and the lower ones ſouth-eaſt. From hence it is, that the corners of theſe ſtrata always ſtrike out at the ſurface into the ſtreets, and cut the ſhoes in pieces. I have likewiſe ſeen ſome ſtrata, inclining to the northward, but nearly perpendicular as the former. Horizontal ſtrata, or nearly ſuch, have occurred to me too. The ſtrata are divided by narrow cracks, which are commonly filled with fibrous white gypſum, which can ſometimes be got looſe with a knife, if the layer or ſtratum of ſlate above it is broken in pieces; and in that caſe it has the appearance of a thin white leaf. The larger cracks are almoſt filled up with tranſparent quartz cryſtals, of different ſizes. One part of the mountain contains vaſt quantities of theſe cryſtals, from which the corner of the mountain which lies to the S. S. E. of the palace, has got the name of *Pointe de Diamante*, or Diamond Point. The ſmall cracks which divide the ſtone, go generally at right angles; the diſtances between them are not always equal. The outſide of the ſtratum, or that which is turned towards the other ſtratum, is frequently covered with a fine, black, ſhining membrane, which looks like a kind of a pyrous horn-ſtone. In it there is sometimes a yellow pyrites, always lying in ſmall grains. I never found petrefactions or impreſſions, or other kinds of ſtone in it, beſides thoſe I have juſt mentioned. The whole mountain on which *Quebec* is ſituated, conſiſts entirely of lime-ſlate from top to bottom. When this ſtone is broken, or ſcraped with a knife, it gives a ſtrong ſmell like the ſtink-ſtone. That part of the mountain which is expoſed to the open air, crumbles into ſmall pieces, had loſt their black colour, and got a pale red one in its ſtead. Almoſt all the public and private buildings at *Quebec* conſiſt of this lime-ſlate; and likewiſe the walls round the town, and round the monaſteries and gardens. It is eaſily broken, and cut to the ſize wanted. But it has the property of ſplitting into thin ſhivers, parallel to the ſurface of the ſtratum from whence they are taken, after lying during one or more years in the air, and expoſed to the ſun. However, this quality does no damage to the walls in which they are placed; for the stones being laid on purpose into such a position that the cracks always run horizontally, the upper stones press so much upon the lower ones, that they can only get cracks outwardly, and shiver only on the outside, without going further inwards. The shivers always grow thinner, as the houses grow older.

In order to give my readers some idea of the climate of *Quebec*, and of the different changes of heat and cold, at the several seasons of the year, I will here insert some particulars extracted from the meteorological observations, of the royal physician, Mr. *Gaulthier:* he gave me a copy of those which he had made from *October* 1744, to the end of *September* 1746. The thermometrical observations I will omit, because I do not think them accurate; for as Mr. *Gaulthier* made use of *de la Hire*'s thermometer, the degrees of cold cannot be exactly determined, the quicksilver being depressed into the globe at the bottom, as soon as the cold begins to be considerable. The observations are made throughout the year, between seven and eight in the morning, and two and three in the afternoon. He has seldom made any observations in the afternoon. His thermometer

ter was likewise inaccurate, by being placed in a bad situation.

The year 1745.

January. THE 29th of this month the river St. *Lawrence* was covered over with ice, near *Quebec*. In the observations of other years, it is observed, that the river is sometimes covered with ice in the beginning of *January,* or the end of *December.*

February. NOTHING remarkable happend during the course of this month.

March. THEY say this has been the mildest winter they ever felt; even the eldest persons could not remember one so mild. The snow was only two feet deep, and the ice in the river, opposite *Quebec,* had the same thickness. On the twenty-first there was a thunder-storm, which fell upon a soldier, and hurt him very much. On the 19th and 20th, they began to make incisions into the sugar-maple, and to prepare sugar from its juice.

April. DURING this month they continued to extract the juice of the sugar-maple, for making sugar. On the 7th the gardeners began to make hot-beds. On the 20th the ice in the river broke loose near *Quebec,* and went down; which rarely happens so soon; for the river St.

Lawrence is sometimes covered with ice opposite *Quebec*, on the 10th of *May*. On the 22d, and 23d, there fell a quantity of snow. On the 25th they began to sow near St. *Joachim*. The same day they saw some swallows. The 29th they sowed corn all over the country. Ever since the 23d the river had been clear at *Quebec*.

May. The third of this month the cold was so great in the morning, that *Celsius*'s, or the *Swedish* thermometer, was four degrees below the freezing point; however, it did not hurt the corn. On the 16th all the summer-corn was sown. On the 5th the *Sanguinaria, Narcissus,* and violet, began to blow. The 17th the wild cherry-trees, rasberry-bushes, apple-trees, and lime-trees, began to expand their leaves. The strawberries were in flower about that time. The 29th the wild cherry-trees were in blossom. On the 26th part of the *French* apple-trees, cherry-trees, and plum-trees, opened their flowers.

June. The 5th of this month all the trees had got leaves. The apple-trees were in full flower. Ripe straw-berries were to be had on the 22d. Here it is noted, that the weather was very fine for the growth of vegetables.

July. The corn began to shoot into ears on the 12th, and had ears every where on the 21ft. (It is to be obferved, that they fow nothing but fummer-corn here.) Soon after the corn began to flower. Hay making began the 22d. All this month the weather was excellent.

Auguft. On the 12th there were ripe pears and melons at *Montreal.* On the 20th the corn was ripe round *Montreal,* and the harveft was begun there. On the 22d the harveft began at *Quebec.* On the 30th, and 31ft, there was a very fmall hoar-froft on the ground.

September. The harveft of all kinds of corn ended on the 24th, and 25th. Melons, water-melons, cucumbers, and fine plums, were very plentiful during the courfe of this month. Apples and pears were likewife ripe, which is not always the cafe. On the laft days of this month they began to plough the land. The following is one of the obfervations of this month: " The old people in this country
" fay, that the corn was formerly never
" ripe till the 15th, or 16th, of *September,*
" and fometimes on the 12th; but no
" fooner. They likewife aflert, that it
" never was perfectly ripe. But fince the
" woods have been fufficiently cleared, the
" beams of the fun have had more room
" to operate, and the corn ripens fooner
" than

" than before *." It is further remarked, that the hot summers are always very fruitful

* It is not only the clearing of woods, but cultivation, and population, that alter the climate of a country, and make it mild. The *Romans* looked upon the winters of *Germany* and *England* as very severe, but happily both countries have at present a much more mild climate than formerly, owing to the three above mentioned reasons. Near *Petersburg*, under sixty degrees north latitude, the river *Neva* was covered with ice 1765, in the beginning of *December*, and cleared of it *April* the 11th 1766. At *Tsaritsin*, which is under forty-eight degrees forty minutes north latitude, the river *Volga* was covered with ice the 26th of *November* 1765, and the ice broke in the river *April* the 27th 1766, (all old stile). Is it not almost incredible, that in a place very near twelve degrees more to the south, the effects of cold should be felt longer, and more severely, than in the more northern climate. And though the neighbourhood of *Petersburg* has a great many woods, the cold was, however, less severe, and lasting; *Tsaritsin* on the contrary has no woods for many hundred miles in its neighbourhood, if we except some few trees and bushes, along the *Volga* and its isles, and the low land along it. Wherever the eye looks to the east, there are vast plains without woods, for many hundred miles. The clearing a country of woods, cannot therefore alone contribute so much to make the climate milder, But cultivation does more. On a ploughed field the snow will always sooner melt, than on a field covered with grass. The inflammable warm perticles brought into the field, by the various kinds of manure, contribute much to soften the rigours of the climate; but the exhalations of thousands of men and cattle, in a populous country, the burning of so many combustibles, and the dispersion of so many caustic particles, through the whole athmosphere; these are things which contribute so much towards softening the rigours of a climate. In a hundred square miles near *Tsaritsin*, there is not so much cultivated land as there is within ten near *Petersburg*; it is in proportion to the number of the inhabitants of both places, and

ful in *Canada*, and that moſt of the corn has hardly ever arrived at perfect maturity.

October. During this month the fields were ploughed, and the weather was very fine all the time. There was a little froſt for ſeveral nights, and on the 28th it ſnowed. Towards the end of this month the trees began to ſhed their leaves.

November. They continued to plough till the 10th of this month, when the trees had ſhed all their leaves. Till the 18th the cattle went out of doors, a few days excepted, when bad weather had kept them at home. On the 16th there was ſome thunder and lightning. There was not yet any ice in the river St. *Lawrence* on the 24th.

December. During this month it is obſerved, that the autumn has been much milder than uſual. On the 1ſt a ſhip could ſtill ſet ſail for *France*; but on the 16th the river St. *Lawrence* was covered with ice on the ſides, but open in the middle.

In

---

and this makes the chief difference of the climate. There is ſtill another conſideration, *Peterſburg* lies near the ſea, and *Tſaritſin* in an inland country; and, generally ſpeaking, countries near the ſea have been obſerved to enjoy a milder climate. Theſe few remarks will be, I believe, ſufficient to enable every body to judge of the changes of the climate in various countries, which, no doubt, grow warmer and more temperate, as cultivation and population increaſe. F.

In the river *Charles* the ice was thick enough for horses with heavy loads to pass over it. On the 26th the ice in the river St. *Lawrence* was washed away by a heavy rain; but on the 28th part of that river was again covered with ice.

THE next observations shew, that the winter has likewise been one of the mildest. I now resume the account of my own journey.

THIS evening I left *Quebec* with a fair wind. The governor-general of *Canada*, the marquis *de la Jonquiere*, ordered one of the king's boats, and seven men to bring me to *Montreal*. The middle of the boat was covered with blue cloth, under which we were secured from the rain. This journey I made at the expence of the *French* king. We went three *French* miles to-day.

September the 12th. WE continued our journey during all this day.

THE small kind of maize, which ripens in three months time, was ripe about this time, and the people drew it out of the ground, and hung it up to dry.

THE weather about this time was like the beginning of our *August*, old stile. Therefore it seems, autumn commences a whole month later in *Canada*, than in the midst of *Sweden*.

NEAR

Near each farm there is a kitchen-garden, in which onions are moſt abundant; becauſe the *French* farmers eat their dinners of them with bread, on Fridays and Saturdays, or faſting days. However, I cannot ſay, the *French* are ſtrict obſervers of faſting; for ſeveral of my rowers ate fleſh to-day, though it was Friday. The common people in *Canada* may be ſmelled when one paſſes by them, on account of their frequent uſe of onions. Pumpions are likewiſe abundant in the farmer's gardens. They dreſs them in ſeveral ways, but the moſt common is to cut them through the middle, and place the inſide on the hearth, towards the fire, till it is quite roaſted. The pulp is then cut out of the peel, and eaten; people above the vulgar put ſugar to it. Carrots, ſallad, *French* beans, cucumbers, and currant ſhrubs, are planted in every farmer's little kitchen-garden.

Every farmer plants a quantity of tobacco near his houſe, in proportion to the ſize of his family. It is likewiſe very neceſſary that they ſhould plant tobacco, becauſe it is ſo univerſally ſmoaked by the common people. Boys of ten or twelve years of age, run about with the pipe in their mouths, as well as the old people.

Perfons above the vulgar, do not refufe to fmoak a pipe now and then. In the northern parts of *Canada*, they generally fmoak tobacco by itfelf; but further upwards, and about *Montreal*, they take the inner bark of the red Cornelian cherry *, crufh it, and mix it with the tobacco, to make it weaker. People of both fexes, and of all ranks, ufe fnuff very much. Almoft all the tobacco, which is confumed here, is the produce of the country, and fome people prefer it even to *Virginian* tobacco: but thofe who pretend to be connoiffeurs, reckon the laft kind better than the other.

THOUGH many nations imitate the *French* cuftoms; yet I obferved on the contrary, that the *French* in *Canada* in many refpects follow the cuftoms of the *Indians*, with whom they converfe every day. They make ufe of the tobacco-pipes, fhoes, garters, and girdles, of the *Indians*. They follow the *Indian* way of making war with exactnefs; they mix the fame things with tobacco; they make ufe of the *Indian* bark-boats, and row them in the *Indian* way; they wrap fquare pieces of cloth round their feet, inftead of ftockings, and have adopted many other *Indian* fafhions. When

* *Cornus fanguinea*, Linn.

one

one comes into the house of a *Canada* peasant, or farmer, he gets up, takes his hat off to the stranger, desires him to sit down, puts his hat on and sits down again. The gentlemen and ladies, as well as the poorest peasants and their wives, are called *Monsieur* and *Madame*. The peasants, and especially their wives, wear shoes, which consist of a piece of wood hollowed out, and are made almost as slippers. Their boys, and the old peasants themselves, wear their hair behind in a cue; and most of them wear red woollen caps at home, and sometimes on their journies.

The farmers prepare most of their dishes of milk. Butter is but seldom seen, and what they have is made of sour cream, and therefore not so good as *English* butter. Many of the *French* are very fond of milk, which they eat chiefly on fasting days. However, they have not so many methods of preparing it as we have in *Sweden*. The common way was to boil it, and put bits of bread, and a good deal of sugar, into it. The *French* here eat near as much flesh as the *English*, on those days when their religion allows it. For excepting the soup, the sallads, and the desert, all their other dishes consist of flesh variously prepared.

AT

AT night we lay at a farm-houfe, near a river called *Petite Riviere*, which falls here into the river St. *Lawrence*. This place is reckoned fixteen *French* miles from *Quebec*, and ten from *Trois Rivieres*. The tide is ftill confiderable here. Here is the laft place where the hills, along the river, confift of black lime-flate; further on they are compofed merely of earth.

FIRE-FLIES flew about the woods at night, though not in great numbers; the *French* call them *Mouches à feu*.

THE houfes in this neighbourhood are all made of wood. The rooms are pretty large. The inner roof refts on two, three, or four, large thick fpars, according to the fize of the room. The chinks are filled with clay, inftead of mofs. The windows are made entirely of paper. The chimney is erected in the middle of the room; that part of the room which is oppofite the fire, is the kitchen; that which is behind the chimney, ferves the people to fleep, and receive ftrangers in. Sometimes there is an iron ftove behind the chimney.

September the 13th. NEAR *Champlain*, which is a place about five *French* miles from *Trois Rivieres*, the fteep hills near the river confift of a yellow, and fometimes ockre-coloured fandy earth, in which
a num-

a number of small springs arise. The water in them is generally filled with yellow ockre, which is a sign, that these dry sandy fields contain a great quantity of the same iron ore, which is dug at *Trois Rivieres*. It is not conceivable from whence that number of small rivulets takes their rise, the ground above being flat, and exceeding dry in summer. The lands near the river are cultivated for about an *English* mile into the country; but behind them there are thick forests, and low grounds. The woods, which collect a quantity of moisture, and prevent the evaporation of the water, force it to make its way under ground to the river. The shores of the river are here covered with a great deal of black iron-sand.

Towards evening we arrived at *Trois Rivieres*, where we staid no longer, than was necessary to deliver the letters, which we brought with us from *Quebec*. After that we went a *French* mile higher up, before we took our night's lodging.

This afternoon we saw three remarkable old people. One was an old Jesuit, called father *Joseph Aubery*, who had been a missionary to the converted *Indians* of St. *François*. This summer he ended the fiftieth year of his mission. He therefore returned to *Quebec*, to renew his vows there; and he seemed to be healthy, and in good spirits. The other two people were our landlord and his wife; he was above eighty years of age, and she was not much younger. They had now been fifty-one years married. The year before, at the end of the fiftieth year of their marriage, they went to church together, and offered up thanks to God Almighty for the great grace he gave them. They were yet quite well, content, merry, and talkative. The old man said, that he was at *Quebec* when the *English* besieged it, in the year 1690, and that the bishop went up and down the streets, dressed in his pontifical robes, and a sword in his hand, in order to recruit the spirits of the soldiers.

THIS old man said, that he thought the winters were formerly much colder than they are now. There fell likewise a greater quantity of snow, when he was young. He could remember the time when pumpions, cucumbers, &c. were killed by the frost about mid-summer, and he assured me, that the summers were warmer now than they used to be formerly. About thirty and some odd years ago, there was such a severe winter in *Canada*, that the frost killed many birds; but the old man could

could not remember the particular year. Every body allowed, that the summers in 1748, and 1749, had been warmer in *Canada* than they have been many years ago.

The soil is reckoned pretty fertile; and wheat yields nine or ten grains from one. But when this old man was a boy, and the country was new and rich every where, they could get twenty, or four-and-twenty, grains from one. They sow but little rye here; nor do they sow much barley, except for the use of cattle. They complain, however, that when they have a bad crop, they are obliged to bake bread of barley.

September the 14th. This morning we got up early, and pursued our journey. After we had gone about two *French* miles, we got into lake St. *Pierre*, which we crossed. Many plants, which are common in our *Swedish* lakes, swim at the top of this water. This lake is said to be covered every winter with such strong ice, that a hundred loaded horses could go over it together with safety.

A CRAW-FISH, or river lobster, somewhat like a crab, but quite minute, about two geometrical lines long, and broad in proportion, was frequently drawn up by us with the aquatic weeds. Its colour is a pale greenish white.

The cordated *Pontederia* * grows plentiful on the sides of a long and narrow canal of water, in the places frequented by our water-lilies †. A great number of hogs wade far into this kind of strait, and sometimes duck the greatest part of their bodies under water, in order to get at the roots, which they are very fond of.

As soon as we were got through lake St. *Pierre*, the face of the country was entirely changed, and became as agreeable as could be wished. The isles, and the land on both sides of us, looked like the prettiest pleasure-gardens; and this continued till near *Montreal*.

Near every farm on the river-side there are some boats, hollowed out of the trunks of single trees, but commonly neat and well made, having the proper shape of boats. In one single place I saw a boat made of the bark of trees.

September the 15th. We continued our journey early this morning. On account of the strength of the river, which came down against us, we were sometimes obliged to let the rowers go on shore, and draw the boat.

* *Pontederia cordata* Linn.
† *Nymphæa*.

At

At four o'clock in the evening we arrived at *Montreal*; and our voyage was reckoned a happy one, becaufe the violence of the river flowing againft us all the way, and the changeablenefs of the winds, commonly protract it to two weeks.

September the 19th. Several people here in town have got the *French* vines, and planted them in their gardens. They have two kinds of grapes, one of a pale green, or almoft white; the other, of a reddifh brown colour. From the white ones they fay, white wine is made; and from the red ones, red wine. The cold in winter obliges them to put dung round the roots of the vines, without which they would be killed by the froft. The grapes began to be ripe in thefe days; the white ones are a little fooner ripe than the red ones. They make no wine of them here, becaufe it is not worth while; but they are ferved up at deferts. They fay thefe grapes do not grow fo big here as in *France*.

Water-melons * are cultivated in great plenty in the *Englifh* and *French American* colonies; and there is hardly a peafant here, who has not a field planted with them. They are chiefly cultivated in the

* *Cucurbita citrullus* Linn.

neighbourhood of towns; and they are very rare in the north part of *Canada*. The *Indians* plant great quantities of water-melons at prefent; but whether they have done it of old is not eafily determined. For an old *Onidoe Indian* (of the fix *Iroquefe* nations) affured me, that the *Indians* did not know water-melons before the *Europeans* came into the country, and communicated them to the *Indians*. The *French*, on the other hand, have affured me, that the *Illinois Indians* have had abundance of this fruit, when the *French* firft came to them; and that they declare, they had planted them fince times immemorial. However, I do not remember having read that the *Europeans*, who firft came to *North-America*, mention the water-melons, in fpeaking of the difhes of the *Indians* at that time. How great the fummer heat is in thofe parts of *America* which I have paffed through, can eafily be conceived, when one confiders, that in all thofe places, they never fow water-melons in hot-beds, but in the open fields in fpring, without fo much as covering them, and they ripen in time. Here are two fpecies of them, viz. one with a red pulp, and one with a white one. The firft is more common to the fouthward, with the *Illinois*, and in the

Englifh

English colonies; the last is more abundant in *Canada*. The seeds are sown in spring, after the cold is entirely gone off, in a good rich ground, at some distance from each other; because their stalks spread far, and require much room, if they shall be very fruitful. They were now ripe at *Montreal*; but in the *English* colonies they ripen in *July* and *August*. They commonly require less time to ripen in, than the common melons. Those in the *English* colonies are commonly sweeter, and more agreeable, than the *Canada* ones. Does the greater heat contribute any thing towards making them more palatable? Those in the province of *New-York* are, however, reckoned the best.

THE water-melons are very juicy; and the juice is mixed with a cooling pulp, which is very good in the hot summer-season. Nobody in *Canada*, in *Albany*, and in other parts of *New-York*, could produce an example, that the eating of water-melons in great quantities had hurt any body; and there are examples even of sick persons eating them without any danger. Further to the south, the frequent use of them it is thought brings on intermitting fevers, and other bad distempers, especially in such people as are less used to them. Many

Frenchmen assured me, that when people born in *Canada* came to the *Illinois*, and eat several times of the water-melons of that part, they immediately got a fever; and therefore the *Illinois* advise the *French* not to eat of a fruit so dangerous to them. They themselves are subject to be attacked by fevers, if they cool their stomachs too often with water-melons. In *Canada* they keep them in a room, which is a little heated; by which means they will keep fresh two months after they are ripe; but care must be taken, that the frost spoil them not. In the *English* plantations they likewise keep them fresh in dry cellars, during part of the winter. They assured me that they keep better when they are carefully broke off from the stalk, and afterwards burnt with a red-hot iron, in the place where the stalk was fastened. In this manner they may be eaten at *Christmas*, and after. In *Pensylvania*, where they have a dry sandy earth, they make a hole in the ground, put the water-melons carefully into it with their stalks, by which means they keep very fresh during a great part of winter. Few people, however, take this trouble with the water-melons; because they being very cooling, and the winter being very cold too, it seems to be less necessary to keep

keep them for eating in that season, which is already very cold. They are of opinion in these parts, that cucumbers cool more than water-melons. The latter are very strongly diuretic. The *Iroquese* call them *Onoheserakatee*.

GOURDS of several kinds, oblong, round, flat or compressed, crook-necked, small, &c. are planted in all the *English* and *French* colonies. In *Canada*, they fill the chief part of the farmers kitchen-gardens, though the onions came very near up with them. Each farmer in the *English* plantations, has a large field planted with gourds, and the *Germans, Swedes, Dutch,* and other *Europeans,* settled in their colonies, plant them. Gourds are a considerable part of the *Indian* food; however, they plant more squashes than common gourds. They declare, that they have had gourds long before the *Europeans* discovered *America*; which seems to be confirmed by the accounts of the first *Europeans* that came into these parts, who mentioned gourds as common food among the *Indians*. The *French* here call them *citrouilles*, and the *English* in the colonies, *pumpkins*. They are planted in spring, when they have nothing to fear from the frost, in an enclosed field, and a good rich soil. They are likewise frequently put into old hot-

hot-beds. In *Canada*, they ripen towards the beginning of *September*, but further southward they are ripe at the end of *July*. As soon as the cold weather commences, they take off all the pumpions that remain on the stalk, whether ripe or not, and spread them on the floor, in a part of the house, where the unripe ones grow perfectly ripe, if they are not laid one upon the other. This is done round *Montreal* in the middle of *September*; but in *Penfylvauia*, I have seen some in the fields on the 19th of *October*. They keep fresh for several months, and even throughout the winter, if they be well secured in dry cellars (for in damp ones they rot very soon) where the cold cannot come in, or, which is still better, in dry rooms which are heated now and then, to prevent the cold from damaging the fruit.

PUMPIONS are prepared for eating in various ways. The *Indians* boil them whole, or roast them in ashes, and eat them then, or go to sell them thus prepared in the towns, and they have, indeed, a very fine flavour, when roasted. The *French* and *English* slice them, and put the slices before the fire to roast; when they are roasted, they generally put sugar on the pulp. Another way of roasting them, is to cut them through the middle, take out all the seeds, put the halves together again, and roast them in an oven.

oven. When they are quite roasted, some butter is put in, whilst they are warm, which being imbibed into the pulp, renders it very palatable. They often boil pumpions in water, and afterwards eat them, either alone or with flesh. Some make a thin kind of pottage of them, by boiling them in water, and afterwards macerating the pulp. This is again boiled with a little of the water, and a good deal of milk, and stirred about whilst it is boiling. Sometimes the pulp is stamped and kneaded into dough, with maize flour or other flour; of this they make cakes. Some make puddings and tarts of gourds. The *Indians*, in order to preserve the pumpions for a very long time, cut them in long slices, which they fasten or twist together, and dry them either by the sun, or by the fire in a room. When they are thus dried, they will keep for years together, and when boiled, they taste very well. The *Indians* prepare them thus at home and on their journies, and from them the *Europeans* have adopted this method. Sometimes they do not take the time to boil it, but eat it dry with hung beef, or other flesh; and I own they are eatable in that state, and very welcome to a hungry stomach. They sometimes preserve them in the following manner at *Montreal*: They cut

cut a pumpion in four pieces, peal them, and take the feeds out of them. The pulp is put in a pot with boiling water, in which it muſt boil from four to ſix minutes. It is then put into a cullender, and left in it till the next day, that the water may run off. When it is mixed with cloves, cinnamon, and ſome lemon peel, preſerved in ſyrup, and there muſt be an equal quantity of ſyrup and of the pulp. After which it is boiled together, till the ſyrup is entirely imbibed, and the white colour of the pulp is quite loſt.

September the 20th. THE corn of this year's harveſt in *Canada*, was reckoned the fineſt they had ever had. In the province of *New-York*, on the contrary, the crop was very poor. The autumn was very fine this year in *Canada*.

September the 22d. THE *French* in *Canada* carry on a great trade with the *Indians*; and though it was formerly the only trade of this extenſive country, yet its inhabitants were conſiderably enriched by it. At preſent, they have beſides the *Indian* goods, ſeveral other articles which are exported from hence. The *Indians* in this neighbourhood, who go hunting in winter like the other *Indians* nations, commonly bring their furs and ſkins to ſale in the neighbouring
French

French towns; however this is not sufficient. The *Indians* who live at a greater diſtance, never come to *Canada* at all; and, leſt they ſhould bring their goods to the *Engliſh*, as the *Engliſh* go to them, the *French* are obliged to undertake journies, and purchaſe the *Indian* goods in the country of the *Indians*. This trade is chiefly carried on at *Montreal*, and a great number of young and old men every year, undertake long and troubleſome voyages for that purpoſe, carrying with them ſuch goods as they know the *Indians* like, and are in want of. It is not neceſſary to take money on ſuch a journey, as the *Indians* do not value it; and indeed I think the *French*, who go on theſe journies, ſcarce ever take a ſol or penny with them.

I will now enumerate the chief goods which the *French* carry with them for this trade, and which have a good run among the *Indians*.

Muſkets, *Powder*, *Shot*, and *Balls*. The *Europeans* have taught the *Indians* in their neighbourhood the uſe of fire-arms, and they have laid aſide their bows and arrows, which were formerly their only arms, and make uſe of muſkets. If the *Europeans* ſhould now refuſe to ſupply the *Indians* with muſkets, they would be ſtarved to death;

as almost all their food consists of the flesh of the animals, which they hunt; or they would be irritated to such a degree as to attack the *Europeans*. The *Indians* have hitherto never tried to make muskets or similar fire-arms; and their great indolence does not even allow them to mend those muskets which they have got. They leave this entirely to the *Europeans*. As the *Europeans* came into *North-America*, they were very careful not to give the *Indians* any fire-arms. But in the wars between the *French* and *English*, each party gave their *Indian* allies fire-arms, in order to weaken the force of the enemy. The *French* lay the blame upon the *Dutch* settlers in *Albany*, saying, that they began, in 1642, to give their *Indians* fire-arms, and taught them the use of them, in order to weaken the *French*. The inhabitants of *Albany*, on the contrary, assert, that the *French* first introduced this custom, as they would have been too weak to resist the combined force of the *Dutch* and *English* in the colonies. Be this as it will, it is certain that the *Indians* buy muskets from the *Europeans*, and know at present better how to make use of them, than some of their teachers. It is likewise certain, that the *Europeans* gain

considerably

considerably by their trade in muskets and ammunition.

Pieces of white cloth, or of a coarse uncut cloth. The *Indians* constantly wear such pieces of cloth, wrapping them round their bodies. Sometimes they hang them over their shoulders; in warm weather, they fasten them round the middle; and in cold weather, they put them over the head. Both their men and women wear these pieces of cloth, which have commonly several blue or red stripes on the edge.

Blue or red cloth. Of this the *Indian* women make their petticoats, which reach only to their knees. They generally chuse the blue colour.

Shirts and shifts of linen. As soon as an *Indian* fellow, or one of their women, have put on a shirt, they never wash it, or strip it off, till it is entirely torn in pieces.

Pieces of cloth, which they wrap round their legs instead of stockings, like the *Russians*.

Hatchets, knives, scissars, needles, and a steel to strike fire with. These instruments are now common among the *Indians*. They all take these instruments from the *Europeans*, and reckon the hatchets and knives much better, than those which they formerly made of stones and bones. The stone

stone hatchets of the ancient *Indians* are very rare in *Canada*.

Kettles of copper or brass, sometimes tinned in the inside. In these the *Indians* now boil all their meat, and they have a very great run with them. They formerly made use of earthen or wooden pots, into which they poured water, or whatever else they wanted to boil, and threw in red hot stones to make it boil. They do not want iron boilers, because they cannot be easily carried on their continual journies, and would not bear such falls and knocks as their kettles are subject to.

Ear-rings of different sizes, commonly of brass, and sometimes of tin. They are worn by both men and women, though the use of them is not general.

Vermillion. With this they paint their face, shirt, and several parts of the body. They formerly made use of a reddish earth, which is to be found in the country; but, as the *Europeans* brought them vermillion, they thought nothing was comparable to it in colour. Many persons have told me, that they had heard their fathers mention, that the first *Frenchmen* who came over here, got a great heap of furs from the *Indians*, for three times as much cinnabar as would ly on the tip of a knife.

Verdi-

Verdigreafe, to paint their faces green. For the black colour, they make ufe of the foot at the bottom of their kettles, and daub their whole face with it.

Looking glaſſes. The *Indians* are very much pleafed with them, and make ufe of them chiefly when they want to paint themfelves. The men conftantly carry their looking glaffes with them on all their journies; but the women do not. The men, upon the whole, are more fond of dreffing than the women.

Burning glaſſes. Thefe are excellent pieces of furniture in the opinion of the *Indians*; becaufe they ferve to light the pipe without any trouble, which an indolent *Indian* is very fond of.

Tobacco is bought by the northern *Indians*, in whofe country it will not grow. The fouthern *Indians* always plant as much of it as they want for their own confumption. Tobacco has a great run amongft the northern *Indians*, and it has been obferved, that the further they live to the northward, the more they fmoke of tobacco.

Wampum, or, as they are here called, *porcelanes*. They are made of a particular kind of fhells, and turned into little fhort cylindrical beads, and ferve the *Indians* for money and ornament.

Glass beads, of a small size, and white or other colours. The *Indian* women know how to fasten them in their ribbands, pouches, and clothes.

Brass and steel wire, for several kinds of work.

Brandy, which the *Indians* value above all other goods that can be brought them; nor have they any thing, though ever so dear to them, which they would not give away for this liquor. But, on account of the many irregularities which are caused by the use of brandy, the sale of it has been prohibited under severe penalties; however, they do not always pay an implicit obedience to this order.

These are the chief goods which the *French* carry to the *Indians*, and they have a good run among them.

The goods which they bring back from the *Indians*, consist entirely in furs. The *French* get them in exchange for their goods, together with all the necessary provisions they want on the journey. The furs are of two kinds; the best are the northern ones, and the worst sort those from the south.

In the northern parts of *America* there are chiefly the following skins of animals: bears, beavers, elks*, rein-deer †, wolf-lynxes ‡, and martens. They sometimes get martens skins from the south, but they are red, and good for little. *Pichou du Nord* is perhaps the animal which the *English*, near *Hudson*'s bay, call the *wolverene*. To the northern furs belong the bears, which are but few, and foxes, which are not very numerous, and generally black; and several other skins.

The skins of the southern parts are chiefly taken from the following animals: wild cattle, stags, roebucks, otters, *Pichoux du Sud*, of which P. *Charlevoix* makes mention §, and are probably a species of cat-lynx, or perhaps a kind of panther; foxes of various kinds, raccoons, cat-lynxes, and several others.

It is inconceivable what hardships the people in *Canada* must undergo on their journies. Sometimes they must carry their goods a great way by land; frequently they are abused by the *Indians*, and sometimes they are killed by them. They often suffer hunger, thirst, heat, and cold, and are bit by gnats, and exposed to the bites of poi-

* Orignacs.
† Cariboux.
‡ Loup cerviers.
§ In his Hist. de la Nouv. France, Tom. V. p. 158.

fonous fnakes, and other dangerous animals and infects. These deftroy a great part of the youth in *Canada*, and prevent the people from growing old. By this means, however, they become fuch brave foldiers, and fo inured to fatigue, that none of them fear danger or hardfhips. Many of them fettle among the *Indians* far from *Canada*, marry *Indian* women, and never come back again.

The prices of the skins in *Canada*, in the year 1749, were communicated to me by *M. de Couagne*, a merchant at *Montreal*, with whom I lodged. They were as follows:

Great and middle fized *bear* skins, coft five livres.

Skins of *young bears*, fifty fols.

———————— *lynxs*, 25 fols.

———————— *pichoux du fud*, 35 fols.

———————— *foxes* from the fouthern parts, 35 fols.

———————— *otters*, 5 livres.

———————— *raccoons*, 5 livres.

———————— *martens*, 45 fols.

———————— *wolf-lynxes**, 4 livres.

———————— *wolves*, 40 fols.

———————— *carcajoux*, an animal which I do not know, 5 livres.

* Loups cerviers.

Montreal.

SKINS of *vifons*, a kind of martens, which live in the water, 25 fols.

RAW fkins of *elks* *, 10 livres.

——————— *ftags* †.

BAD fkins of *elks* and *ftags* ‡, 3 livres.

SKINS of *roebucks*, 25, or 30 fols.

——————— *red foxes*, 3 livres.

——————— *beavers*, 3 livres.

I WILL now infert a lift of all the different kinds of fkins, which are to be got in *Canada*, and which are fent from thence to *Europe*. I got it from one of the greateft merchants in *Montreal*. They are as follows:

Prepared roebuck fkins, *chevreuils paffés*.
Unprepared ditto, *chevreuils verts*.
Tanned ditto, *chevreuils tanés*.
Bears, *ours*.
Young bears, *ourfons*.
Otters, *loutres*.
Pecans.
Cats, *chats*.
Wolves, *loup de bois*.
Lynxes, *loups cerviers*.
North pichoux, *pichoux du nord*.

* Originacs verts.
† Cerfs verts.
‡ Originacs et cerfs paffés.

South pichoux, *pichoux du sud.*
Red foxes, *renards rouges.*
Cross foxes, *renards croisés.*
Black foxes, *renards noirs.*
Grey foxes, *renards argentés.*
Southern, or *Virginian* foxes, *renards du sud où de* Virginie.
White foxes, from *Tadoussac, renards blancs de* Tadoussac.
Martens, *martres.*
Visons, or *foutreaux.*
Black squirrels, *ecureuils noirs.*
Raw stags skins, *cerfs verts.*
Prepared ditto, *cerfs passés.*
Raw elks skins, *originals verts.*
Prepared ditto, *originals passés.*
Rein-deer skins, *cariboux.*
Raw hinds skins, *biches verts.*
Prepared ditto, *biches passées.*
Carcajoux.
Musk rats, *rats musques.*
Fat winter beavers, *castors gras d'hiver.*
Ditto summer beavers, *castors gras d'été.*
Dry winter beavers, *castors secs d'hiver.*
Ditto summer beavers, *castors secs d'été.*
Old winter beavers, *castors vieux d'hiver.*
Ditto summer beavers, *castors vieux d'été.*

TO-DAY, I got a piece of native copper from the *Upper Lake.* They find it there
almost

almoſt quite pure; ſo that it does not want melting over again, but is immediately fit for working. Father *Charlevoix* * ſpeaks of it in his Hiſtory of *New-France*. One of the Jeſuits at *Montreal*, who had been at the place where this metal is got, told me, that it is generally found near the mouths of rivers, and that there are pieces of native copper too heavy for a ſingle man to lift up. The *Indians* there ſay, that they formerly found a piece of about ſeven feet long, and near four feet thick, all of pure copper. As it is always found in the ground near the mouths of rivers, it is probable that the ice or water carried it down from a mountain; but, notwithſtanding the careful ſearch that has been made, no place has been found, where the metal lies in any great quantity together.

The head or ſuperior of the prieſts of *Montreal*, gave me a piece of lead-ore to-day. He ſaid it was taken from a place only a few *French* miles from *Montreal*, and it conſiſted of pretty compact, ſhining cubes, of lead ore. I was told by ſeveral perſons here, that furthermore ſouthward in the country, there is a place where they find a great quantity of this lead-ore in the ground. The *In-*

---

* See his Hiſt. de la Nouv. Fr. Tom. VI. p. 415.

dians near it, melt it, and make balls and shot of it. I got some pieces of it likewise, consisting of a shining cubic lead-ore, with narrow stripes between it, and of a white hard earth or clay, which effervesces with *aqua fortis*.

I LIKEWISE received a reddish brown earth to-day, found near the *Lac de Deux Montagnes*, or *Lake of Two Mountains*, a few *French* miles from *Montreal*. It may be easily crumbled into dust between the fingers. It is very heavy, and more so than the earth of that kind generally is. Outwardly, it has a kind of glossy appearance, and, when it is handled by the fingers for some time, they are quite as it were silvered over. It is, therefore, probably a kind of lead-earth or an earth mixed with ironglimmer.

THE ladies in *Canada* are generally of two kinds: some come over from *France*, and the rest natives. The former possess the politeness peculiar to the *French* nation; the latter may be divided into those of *Quebec* and *Montreal*. The first of these are equal to the *French* ladies in good breeding, having the advantage of frequently conversing with the *French* gentlemen and ladies, who come every summer with the king's ships, and stay several weeks

at

Montreal.

at *Quebec*, but feldom go to *Montreal*. The
ladies of this laft place are accufed by the
French of partaking too much of the pride
of the *Indians*, and of being much wanting in *French* good breeding. What I
have mentioned above of their dreffing
their head too affiduoufly, is the cafe with
all the ladies throughout *Canada*. Their
hair is always curled, even when they are
at home in a dirty jacket, and fhort coarfe
petticoat, that does not reach to the middle of their legs. On thofe days when
they pay or receive vifits, they drefs fo gayly,
that one is almoft induced to think their
parents poffeffed the greateft dignities in
the ftate. The *Frenchmen*, who confidered
things in their true light, complained very
much that a great part of the ladies in *Canada* had got into the pernicious cuftom of
taking too much care of their drefs, and
fquandering all their fortunes, and more,
upon it, inftead of fparing fomething for
future times. They are no lefs attentive
to have the neweft fafhions; and they laugh
at each other, when they are not dreffed to
each other's fancy. But what they get as
new fafhions, are grown old, and laid afide
in *France*; for the fhips coming but once
every year from thence, the people in *Canada* confider that as the new fafhion for

the

the whole year, which the people on board brought with them, or which they impofed upon them as new. The ladies in *Canada*, and efpecially at *Montreal*, are very ready to laugh at any blunders ftrangers make in fpeaking; but they are very excufable. People laugh at what appears uncommon and ridiculous. In *Canada* nobody ever hears the *French* language fpoken by any but *Frenchmen*; for ftrangers feldom come thither; and the *Indians* are naturally too proud to learn *French*, but oblige the *French* to learn their language. From hence it naturally follows, that the nice *Canada* ladies cannot hear any thing uncommon without laughing at it. One of the firft queftions they propofe to a ftranger is, whether he is married? The next, how he likes the ladies in the country; and whether he thinks them handfomer than thofe of his own country? And the third, whether he will take one home with him? There are fome differences between the ladies of *Quebec*, and thofe of *Montreal*; thofe of the laft place feemed to be generally handfomer than thofe of the former. Their behaviour likewife feemed to me to be fomewhat too free at *Quebec*, and of a more becoming modefty at *Montreal*. The ladies at *Quebec*, efpecially the unmarried ones, are not very induftrious. A girl of eighteen

eighteen is reckoned very poorly off, if she cannot enumerate at least twenty lovers. These young ladies, especially those of a higher rank, get up at seven, and dress till nine, drinking their coffee at the same time. When they are dressed, they place themselves near a window that opens into the street, take up some needle-work, and sew a stitch now and then; but turn their eyes into the street most of the time. When a young fellow comes in, whether they are acquainted with him or not, they immediately lay aside their work, sit down by him, and begin to chat, laugh, joke, and invent *double-entendres*; and this is reckoned being very witty*. In this manner they frequently pass the whole day, leaving their mothers to do all the business in the house. In *Montreal*, the girls are not quite so volatile, but more industrious. They are always at their needle-work, or doing some necessary business in the house. They are likewise chearful and content; and nobody can say that they want either wit, or charms. Their fault is, that they think too well of themselves. However, the daughters of people of all ranks, without exception, go to market, and carry home what they have bought. They rise as soon,

* *Avoir beaucoup d'esprit.*

and

and go to bed as late, as any of the people in the house. I have been assured, that, in general, their fortunes are not considerable; which are rendered still more scarce by the number of children, and the small revenues in a house. The girls at *Montreal* are very much displeased that those at *Quebec* get husbands sooner than they. The reason of this is, that many young gentlemen who come over from *France* with the ships, are captivated by the ladies at *Quebec*, and marry them; but as these gentlemen seldom go up to *Montreal*, the girls there are not often so happy as those of the former place.

September the 23d. THIS morning I went to *Saut au Recollet*, a place three *French* miles northward of *Montreal*, to describe the plants and minerals there, and chiefly to collect seeds of various plants. Near the town there are farms on both sides of the road; but as one advances further on, the country grows woody, and varies in regard to height. It is generally very strong; and there are both pieces of rock-stone, and a kind of grey lime-stone. The roads are bad, and almost impassable for chaises. A little before I arrived at *Saut au Recollet*, the woods end, and the country is turned into corn-fields, meadows, and pastures.

ABOUT a *French* mile from the town are two lime-kilns on the road. They are built of a grey lime-ftone, burnt hard, and of pieces of rock-ftone, towards the fire. The height of the kiln from top to bottom is feven yards.

THE lime-ftone which they burn here, is of two kinds. One is quite black, and fo compact, that its conftituent particles cannot be diftinguifhed, fome difperfed grains of white and pale grey fpar excepted. Now and then there are thin cracks in it filled with a white fmall-grained fpar.

I HAVE never feen any petrefactions in this ftone, though I looked very carefully for them. This ftone is common on the ifle of *Montreal,* about ten or twenty inches below the upper foil. It lies in ftrata of five or ten inches thicknefs. This ftone is faid to give the beft lime; for, though it is not fo white as that of the following grey lime-ftone, yet it makes better mortar, and almoft turns into ftone, growing harder and more compact every day. There are examples, that when they have been about to repair a houfe made partly of this mortar, the other ftones of which the houfe confifts, fooner broke in pieces than the mortar itfelf.

The other kind is a grey, and sometimes a dark grey lime-stone, consisting of a compact calcareous-stone, mixed with grains of spar, of the same colour. When broken, it has a strong smell of stink-stone. It is full of petrified striated shells or pectinites. The greatest part of these petrefactions are, however, only impressions of the hollow side of the shells. Now and then I found likewise petrefied pieces of the shell itself, though I could never find the same shells in their natural state on the shores; and it seems inconceivable how such a quantity of impressions could come together, as I shall presently mention.

I have had great pieces of this lime-stone, consisting of little else than pectinites, lying close to one another. This lime-stone is found on several parts of the isle, where it lies in horizontal strata of the thickness of five or ten inches. This stone yields a great quantity of white lime, but it is not so good as the former, because it grows damp in wet weather.

Fir-wood is reckoned the best for the lime-kilns, and the thuya wood next to it. The wood of the sugar-maple, and other trees of a similar nature, are not fit for it, because they leave a great quantity of coals.

Grey pieces of rock-stone are to be seen in the woods and fields hereabouts.

The leaves of several trees and plants began now to get a pale hue; especially those of the red maple, the smooth sumach *, the *Polygonum sagittatum*, Linn. and several of the ferns.

A great cross is erected on the road, and the boy who shewed me the wood, told me that a person was buried there, who had wrought great miracles.

At noon I arrived at *Saut au Recollet*, which is a little place, situated on a branch of the river *St. Lawrence*, which flows with a violent current between the isles of *Montreal* and *Jesus*. It has got its name from an accident which happened to a recollet friar, called *Nicolas Veil*, in the year 1625. He went into a boat with a converted *Indian*, and some *Indians* of the nation of *Hurons*, in order to go to *Quebec*; but, on going over this place in the river, the boat overset, and both the friar and his proselyte were drowned. The *Indians* (who have been suspected of occasioning the oversetting of the boat) swam to the shore, saved what they could of the friars effects, and kept them.

* *Rhus glabrum.* Linn.

THE

The country hereabouts is full of stones, and they have but lately began to cultivate it; for all the old people could remember the places covered with tall woods, which are now turned into corn-fields, meadows, and pastures. The priests say, that this place was formerly inhabited by some converted *Hurons*. These *Indians* lived on a high mountain, at a little distance from *Montreal*, when the *French* first arrived here, and the latter persuaded them to sell that land. They did so, and settled here at *Saut au Recollet*, and the church which still remains here, was built for them, and they have attended divine service in it for many years. As the *French* began to increase on the isle of *Montreal*, they wished to have it entirely to themselves, and persuaded the *Indians* again to sell them this spot, and go to another. The *French* have since prevailed upon the *Indians* (whom they did not like to have amongst them, because of their drunkenness, and rambling idle life) to leave this place again, and go to settle at the lake *des Deux Montagnes*, where they are at present, and have a fine church of stone. Their church at *Saut au Recollet* is of wood, looks very old and ruinous, though its inside is pretty good, and is made use of by the *Frenchmen* in this place. They have already

ready brought a quantity of ſtones hither, and intend building a new church very ſoon. The botanical obſervations which I made during theſe days, I ſhall reſerve for another publication.

Though there had been no rain for ſome days paſt, yet the moiſture in the air was ſo great, that as I ſpread ſome papers on the ground this afternoon, in a ſhady place, intending to put the ſeeds I collected into them, they were ſo wet in a few minutes time, as to be rendered quite uſeleſs. The whole ſky was very clear and bright, and the heat as intolerable as in the middle of *July*.

One half of the corn-fields are left fallow alternately. The fallow grounds are never ploughed in ſummer; ſo the cattle can feed upon the weeds that grow on them. All the corn made uſe of here is ſummer corn, as I have before obſerved. Some plough the fallow grounds late in autumn; others defer that buſineſs till ſpring; but the firſt way is ſaid to give a much better crop. Wheat, barley, rye, and oats are harrowed, but peaſe are ploughed under ground. They ſow commonly about the 15th of *April*, and begin with the peaſe. Among the many kinds of peaſe which are to be got here, they prefer the green ones to all others for sowing. They require a high, dry, poor ground, mixed with coarse sand. The harvest time commences about the end, and sometimes in the middle of *August*. Wheat returns generally fifteen, and sometimes twenty fold; oats from fifteen to thirty fold. The crop of pease is sometimes forty fold, but at other times only ten fold; for they are very different. The plough and harrow are the only instruments of husbandry they have, and those none of the best sort neither. The manure is carried upon the fallow grounds in spring. The soil consists of a grey stony earth, mixed with clay and sand. They sow no more barley than is necessary for the cattle; for they make no malt here. They sow a good deal of oats, but merely for the horses and other cattle. Nobody knows here how to make use of the leaves of deciduous trees as a food for the cattle, though the forests are furnished with no other than trees of that kind, and though the people are commonly forced to feed their cattle at home during five months.

I HAVE already repeatedly mentioned, that almost all the wheat which is sown in *Canada* is summer wheat, that is such as is sown in spring. Near *Quebec* it sometimes happens, when the summer is less warm, or
the

the spring later than common, that a great part of the wheat does not ripen perfectly before the cold commences. I have been assured that some people, who live on the *Isle de Jesus*, sow wheat in autumn, which is better, finer, and gives a more plentiful crop, than the summer wheat; but it does not ripen above a week before the other wheat.

September the 25th. In several places hereabouts, they enclose the fields with a stone fence, instead of wooden pales. The plenty of stones which are to be got here, render the labour very trifling.

Here are abundance of beech trees in the woods, and they now had ripe seeds. The people in *Canada* collect them in autumn, dry them, and keep them till winter, when they eat them, instead of walnuts and hazel nuts; and I am told they taste very well.

There is a salt spring, as the priest of this place informed me, seven *French* miles from hence, near the river *d'Assomption*; of which during the war, they have made a fine white salt. The water is said to be very briny.

Some kinds of fruit-trees succeed very well near *Montreal*, and I had here an opportunity of seeing some very fine pears and apples of various sorts. Near *Quebec* the pear-trees will not succeed, because the winter is too severe for them; and sometimes they are killed by the frost in the neighbourhood of *Montreal*. Plum-trees of several sorts were first brought over from *France*, succeed very well, and withstand the rigours of winter. Three varieties of *America* walnut-trees grow in the woods; but the walnut-trees brought over from *France* die almost every year down to the very root, bringing forth new shoots in spring. Peach-trees cannot well agree with this climate; a few bear the cold, but, for greater safety, they are obliged to put straw round them. Chesnut-trees, mulberry-trees, and the like, have never yet been planted in *Canada*.

The whole cultivated part of *Canada* has been given away by the king to the clergy, and some noblemen; but all the uncultivated parts belong to him, as likewise the place on which *Quebec* and *Trois Rivieres* are built. The ground on which the town of *Montreal* is built, together with the whole isle of that name, belongs to the priests of the order of St. *Sulpicius*, who live at *Montreal*. They have given the land in tenure to farmers and others who were willing to settle on it, in so much that they have more upon their hands at present

present. The first settlers paid a trifling rent for their land; for frequently the whole lease for a piece of ground, three *arpens* broad and thirty long, consists in a couple of chicken; and some pay twenty, thirty, or forty sols for a piece of land of the same size. But those who came later, must pay near two *ecus* (crowns) for such a piece of land, and thus the land-rent is very unequal throughout the country. The revenues of the bishop of *Canada* do not arise from any landed property. The churches are built at the expence of the congregations. The inhabitants of *Canada* do not yet pay any taxes to the king; and he has no other revenues from it, than those which arise from the custom-house.

The priests of *Montreal* have a mill here, where they take the fourth part of all that is ground. However the miller receives a third part of this share. In other places he gets the half of it. The priests sometimes lease the mill for a certain sum. Besides them nobody is allowed to erect a mill on the isle of *Montreal*, they having reserved that right to themselves. In the agreement drawn up between the priests and the inhabitants of the isle, the latter are obliged to get all their corn ground in the mills of the former.

They boil a good deal of sugar in *Canada* of the juice running out of the incisions in the sugar-maple, the red maple, and the sugar-birch; but that of the first tree is most commonly made use of. The way of preparing it has been more minutely described by me, in the Memoirs of the Royal *Swedish* Academy of Sciences *.

September the 26th. Early this morning I returned to *Montreal*. Every thing began now to look like autumn. The leaves of the trees were pale or reddish, and most of the plants had lost their flowers. Those which still preserved them were the following †:

Several sorts of asters, both blue and white.

Golden rods of various kinds.
Common milfoil.
Common self-heal.
The crisped thistle.
The biennial oenothera.
The rough-leaved sun-flower, with trifoliated leaves.
The *Canada* violet.

---

* See the Volume for the year 1751, p. 143, &c.

† *Asteres. Solidagines. Achillea millefolium. Prunella vulgaris. Carduus crispus. Oenothera biennis. Rudbeckia triloba. Viola Canadensis. Gentiana Saponaria.*

A species of gentian.

WILD vines are abundant in the woods hereabouts, climbing up very high trees.

I HAVE made enquiry among the *French*, who travel far into the country, concerning the food of the *Indians*. Those who live far north, I am told, cannot plant any thing, on account of the great degree of cold. They have, therefore, no bread, and do not live on vegetables; flesh and fish is their only food, and chiefly the flesh of beavers, bears, rein-deer, elks, hares, and several kinds of birds. Those *Indians* who live far southward, eat the following things. Of vegetables they plant maize, wild kidney beans * of several kinds, pumpions of different sorts, *squashes*, a kind of gourds, watermelons and melons †. All these plants have been cultivated by the *Indians*, long before the arrival of the *Europeans*. They likewise eat various fruits which grow in their woods. Fish and flesh make a very great part of their food. And they chiefly like the flesh of wild cattle, roe-bucks, stags, bears, beavers, and some other quadrupeds. Among their dainty dishes, they reckon the *water-taregrass* ‡, which the *French* call

* *Phaseoli.*
† *Cucumis melo*, Linn.
‡ *Zizania aquatica*, Linn.

folle

folle avoine, and which grows in plenty in their lakes, in stagnant waters, and sometimes in rivers which flow slowly. They gather its seeds in *October*, and prepare them in different ways, and chiefly as groats, which taste almost as well as rice. They make likewise many a delicious meal of the several kinds of walnuts, chesnuts, mulberries, *acimine**, chinquapins †, hazel-nuts, peaches, wild prunes, grapes, whortle-berries of several sorts, various kinds of medlars, black-berries, and other fruit and roots. But the species of corn so common in what is called the old world, were entirely unknown here before the arrival of the *Europeans*; nor do the *Indians* at present ever attempt to cultivate them, though they see the use which the *Europeans* make of the culture of them, and though they are fond of eating the dishes which are prepared of them.

September the 27th. BEAVERS are abundant all over *North-America*, and they are one of the chief articles of the trade in *Canada*. The *Indians* live upon their flesh during a great part of the year. It is certain that these animals multiply very fast; but it is no less so, that

---
* *Annona muricata*, Linn.
† *Fagus pumila*, Linn.

vaft numbers of them are annually killed, and that the *Indians* are obliged at prefent to undertake diftant journies, in order to catch or fhoot them. Their decreafing in number is very eafily accounted for; becaufe the *Indians*, before the arrival of the *Europeans*, only caught as many as they found neceffary to clothe themfelves with, there being then no trade with the fkins. At prefent a number of fhips go annually to *Europe*, laden chiefly with beavers fkins; the *Englifh* and *French* endeavour to outdo each other, by paying the *Indians* well for them, and this encourages the latter to extirpate thefe animals. All the people in *Canada* told me, that when they were young, all the rivers in the neighbourhood of *Montreal*, the river St. *Lawrence* not excepted, were full of beavers and their dykes; but at prefent they are fo far extirpated, that one is obliged to go feveral miles up the country before one can meet with one. I have already remarked above, that the beaver fkins from the north, are better than thofe from the fouth.

BEAVER-FLESH is eaten not only by the *Indians*, but likewife by the *Europeans*, and efpecially the *French*, on their faiting days; for his holinefs, in his fyftem, has ranged the beaver among the fifh. The fleſh

flesh is reckoned best, if the beaver has lived upon vegetables, such as the asp, and the beaver-tree *; but when he has eaten fish, it does not taste well. To day I tasted this flesh boiled, for the first time; and though every body present, besides myself, thought it a delicious dish, yet I could not agree with them. I think it is eatable, but has nothing delicious. It looks black when boiled, and has a peculiar taste. In order to prepare it well, it must be boiled in several waters from morning till noon, that it may lose the bad taste it has. The tail is likewise eaten, after it has been boiled in the same manner, and roasted afterwards; but it consists of fat only, though they would not call it so; and cannot be swallowed by one who is not used to eat it.

Much has already been written concerning the dykes, or houses of the beavers; it is therefore unnecessary to repeat it. Sometimes, though but seldom, they catch beavers with white hair.

Wine is almost the only liquor which people above the vulgar are used to drink. They make a kind of spruce beer of the top of the white fir †, which they drink

---

* *Magnolia glauca*, Linn.

† *Epinette blanche*. The way of brewing this beer is described at large in the Memoirs of the Royal Acad. of Sciences, for the year 1751, p. 190.

in

in fummer; but the ufe of it is not general; and it is feldom drank by people of quality. Thus great fums go annually out of the country for wine; as they have no vines here, of which they could make a liquor that is fit to be drank. The common people drink water; for it is not yet cuftomary here to brew beer of malt; and there are no orchards large enough to fupply the people with apples for making cyder. Some of the people of rank, who poffefs large orchards, fometimes, out of curiofity, get a fmall quantity of cyder made. The great people here, who are ufed from their youth to drink nothing but wine, are greatly at a lofs in time of war; when all the fhips which brought wine are intercepted by the *Englifh* privateers. Towards the end of the laft war, they gave two hundred and fifty *Francs*, and even one hundred *Ecus*, for a *barrique*, or hogfhead, of wine.

The prefent price of feveral things, I have been told by fome of the greateft merchants here, is as follows. A middling horfe cofts forty *Francs* * and upwards; a good horfe is valued at an hundred *Francs*,

---

* *Franc* is the fame as *Livre*; and twenty-two *Livres* make a pound fterling.

or more. A cow is now sold for fifty
Francs; but people can remember the time
when they were sold for ten *Ecus* *. A
sheep costs five or six livres at present; but
last year, when every thing was dear, it
cost eight or ten *Francs*. A hog of one
year old, and two hundred, or an hundred
and fifty pound weight, is sold at fifteen
Francs. M. *Couagne*, the merchant, told
me, that he had seen a hog of four hundred weight among the *Indians*. A chick-
en is sold for ten or twelve *Sols* †; and a
turkey for twenty sols. A *Minot* ‡ of
wheat sold for an *Ecu* last year; but at
present it cost forty *Sols*. Maize is always
of the same price with wheat, because here
is but little of it; and it is all made use of
by those who go to trade with the *Indians*.
A *Minot* of oats costs sometimes from fifteen to twenty *Sols*; but of late years it
has been sold for twenty-six, or thirty *Sols*.
Pease bear always the same price with
wheat. A pound of butter costs commonly
about eight or ten *Sols*; but last year it rose
up to sixteen *Sols*. A dozen of eggs used
to cost but three *Sols*; however, now are

* An *Ecu* is three *Francs*.
† Twenty *Sols* make one *Livre*.
‡ A *French* measure, about the same as two bushels in *England*.

sold

fold for five. They make no cheese at *Montreal*; nor is there any to be had, except what is got from abroad. A watermelon generally costs five or six *Sols*; but if of a large size, from fifteen to twenty.

THERE are as yet no manufactures established in *Canada*; probably, because *France* will not lose the advantage of selling off its own goods here. However, both the inhabitants of *Canada*, and the *Indians*, are very ill off for want of them, in times of war.

THOSE persons who want to be married, must have the consent of their parents. However, the judge may give them leave to marry, if the parents oppose their union, without any valid reason. Likewise, if the man be thirty years of age, and the woman twenty-six, they may marry, without farther waiting for their parents consent.

September the 29th. THIS afternoon I went out of town, to the south-west part of the isle, in order to view the country, and the œconomy of the people, and to collect several seeds. Just before the town are some fine fields, which were formerly cultivated, but now serve as pastures. To the north-west appears the high mountain, which lies westward of *Montreal*, and is very fertile, and covered with fields and

gardens

gardens from the bottom to the summit. On the south-east side is the river St. *Lawrence*, which is very broad here; and on its sides are extensive corn-fields and meadows, and fine houses of stone, which look white at a distance. At a great distance south-eastward, appear the two high mountains near fort *Chamblais*, and some others near lake *Champlain*, raising their tops above the woods. All the fields hereabouts are filled with stones of different sizes; and among them, there is now and then a black lime-stone. About a *French* mile from the town, the high road goes along the river, which is on the left-hand; and on the right-hand all the country is cultivated and inhabited. The farm-houses are three, four, or five arpens distant from each other. The hills near the river are generally high and pretty steep; they consist of earth; and the fields below them are filled with pieces of rock-stone, and of black lime-slate. About two *French* miles from *Montreal*, the river runs very rapidly, and is full of stones; in some places there are some waves. However, those who go in boats into the southern parts of *Canada*, are obliged to work through such places.

Most of the farm-houses in this neighbourhood are of stone, partly of the black lime-

lime-stone, and partly of other stones in the neighbourhood. The roof is made of shingles or of straw. The gable is always very high and steep. Other buildings, such as barns and stables, are of wood.

WILD-GEESE and ducks, began now to migrate in great flocks to the southern countries.

October the 2d. THE two preceding days, and this, I employed chiefly in collecting seeds.

THE last night's frost had caused a great alteration in several trees. Walnut-trees of all sorts shed their leaves in plenty now. The flowers of a kind of nettle * were all entirely killed by the frost. The leaves of the *American* lime-tree were likewise damaged. In the kitchen-gardens the leaves of the melons were all killed by the frost. However, the beech, oak, and birch, did not seem to have suffered at all. The fields were all covered with a hoar-frost. The ice in the pools of water was a geometrical line and a half in thickness.

THE biennial oenothera † grows in abundance on open woody hills, and fallow

---

* *Urtica divaricata*, Linn.
† *Oenothera biennis*, Linn.

fields.

fields. An old *Frenchman*, who accompanied me as I was collecting its seeds, could not sufficiently praise its property of healing wounds. The leaves of the plant must be crushed, and then laid on the wound.

Sœurs de Congregation are a kind of religious women, different from nuns. They do not live in a convent, but have houses both in the town and country. They go where they please, and are even allowed to marry, if an opportunity offers; but this, I am told, happens very seldom. In many places in the country, there are two or more of them: they have their house commonly near a church, and generally the parsonage house is on the other side of the church. Their business is to instruct young girls in the Christian religion, to teach them reading, writing, needle-work, and other female accomplishments. People of fortune board their daughters with them for some time. They have their boarding, lodging, beds, instruction, and whatever else they want, upon very reasonable terms. The house where the whole community of these ladies live, and from whence they are sent out into the country, is at *Montreal*. A lady that wants to become incorporated among

among them, must pay a considerable sum of money towards the common stock; and some people reckon it to be four thousand livres. If a person be once received, she is sure of a subsistence during her life-time.

La Chine is a fine village, three *French* miles to the south-east of *Montreal*, but on the same isle, close to the river St. *Lawrence*. The farm-houses ly along the river-side, about four or five *arpens* from each other. Here is a fine church of stone, with a small steeple; and the whole place has a very agreeable situation. Its name is said to have had the following origin. As the unfortunate M. *Salée* was here, who was afterwards murdered by his own countrymen further up in the country, he was very intent upon discovering a shorter road to *China*, by means of the river St. *Lawrence*. He talked of nothing at that time but his new short way to *China*. But as his project of undertaking this journey, in order to make this discovery, was stopped by an accident which happened to him here, and he did not that time come any nearer *China*, this place got its name, as it were, by way of joke.

This evening I returned to *Montreal*.

October the 5th. THE governor-general at *Quebec* is, as I have already mentioned before, the chief commander in *Canada*. Next to him is the intendant at *Quebec*; then follows the governor of *Montreal*, and after him the governor of *Trois Rivieres*. The intendant has the greatest power next to the governor-general; he pays all the money of government, and is president of the board of finances, and of the court of justice in this country. He is, however, under the governor-general; for if he refuses to do any thing to which he seems obliged by his office, the governor-general can give him orders to do it, which he must obey. He is allowed, however, to appeal to the government in *France*. In each of the capital towns, the governor is the highest person, then the lieutenant-general, next to him a major, and after him the captains. The governor-general gives the first orders in all matters of consequence. When he comes to *Trois Rivieres* and *Montreal*, the power of the governor ceases, because he always commands where he is. The governor-general commonly goes to *Montreal* once every year, and mostly in winter; and during his absence from *Quebec*, the lieutenant-general commands there. When the governor-general dies, or goes to *France*, before a new one is come in his ſtead, the governor of *Montreal* goes to *Quebec* to command in the mean while, leaving the major to command at *Montreal*.

ONE or two of the king's ſhips are annually ſent from *France* to *Canada*, carrying recruits to ſupply the places of thoſe ſoldiers, who either died in the ſervice, or have got leave to ſettle in the country, and turn farmers, or to return to *France*. Almoſt every year they ſend a hundred, or a hundred and fifty people over in this manner. With theſe people they likewiſe ſend over a great number of perſons, who have been found guilty of ſmuggling in *France*. They were formerly condemned to the gallies, but at preſent they ſend them to the colonies, where they are free as ſoon as they arrive, and can chooſe what manner of life they pleaſe, but are never allowed to go out of the country, without the king's ſpecial licence. The king's ſhips likewiſe bring a great quantity of merchandizes which the king has bought, in order to be diſtributed among the *Indians* on certain occaſions. The inhabitants of *Canada* pay very little to the king. In the year 1748, a beginning was, however,

however, made, by laying a duty of three *per cent.* on all the *French* goods imported by the merchants of *Canada*. A regulation was likewise made at that time, that all the furs and skins exported to *France* from hence, should pay a certain duty; but what is carried to the colonies pays nothing. The merchants of all parts of *France* and its colonies, are allowed to send ships with goods to this place; and the *Quebec* merchants are at liberty likewise to send their goods to any place in *France*, and its colonies. But the merchants at *Quebec* have but few ships, because the sailors wages are very high. The towns in *France* which chiefly trade with *Canada*, are *Rochelle* and *Bourdeaux*; next to them are *Marseilles*, *Nantes*, *Havre de Grace*, *St. Malo*, and others. The king's ships which bring goods to this country, come either from *Brest* or from *Rochefort*. The merchants at *Quebec* send flour, wheat, pease, wooden utensils, *&c.* on their own bottoms, to the *French* possessions in the *West-Indies*. The walls round *Montreal* were built in 1738, at the king's expence, on condition the inhabitants should, little by little, pay off the cost to the king. The town at present pays annually 6000 *livres* for them to government, of which 2000 are given

given by the seminary of priests. At *Quebec* the walls have likewise been built at the king's expence, but he did not redemand the expence of the inhabitants, because they had already the duty upon goods to pay as above mentioned. The beaver trade belongs solely to the *Indian* company in *France,* and nobody is allowed to carry it on here, besides the people appointed by that company. Every other fur trade is open to every body. There are several places among the *Indians* far in the country, where the *French* have stores of their goods; and these places they call *les postes*. The king has no other fortresses in *Canada* than *Quebec, Fort Chamblais, Fort St. Jean, Fort St. Frederic,* or *Crownpoint, Montreal, Frontenac,* and *Niagara.* All other places belong to private persons. The king keeps the *Niagara* trade all to himself. Every one who intends to go to trade with the *Indians* must have a licence from the governor-general, for which he must pay a sum according as the place he is going to is more or less advantageous for trade. A merchant who sends out a boat laden with all sorts of goods, and four or five persons with it, is obliged to give five or six hundred livres for the permission; and there are places for which they give a

thousand

thousand livres. Sometimes one cannot buy the licence to go to a certain trading place, because the governor-general has granted, or intends to grant it to some acquaintaince or relation of his. The money arising from the granting of licences, belongs to the governor-general; but it is customary to give half of it to the poor: whether this is always strictly kept to or not, I shall not pretend to determine.

END OF THE THIRD VOLUME.

INDEX.

A.

Acer Negundo, i. 67.
—— *rubrum*, red maple, i. 66.
Achillæa millefolium, iii. 291.
Adiantum pedatum, maiden-hair, iii. 118.
Albany, town of, ii. 255.
———— fort at, ii. 258.
———— houses of, ii. 256.
———— inhabitants of, ii. 21.
———— situation of, ii. 258.
Albecor, i. 19.
Algonkin words, iii. 204.
Allium Canadense? ii. 133.
Anas acuta, blue bill, i. 237.
Anemone hepatica, ii. 104.
Anies, iii. 181.
Animals, tameable, i. 207.
Anne fort near Canada, ii. 297.
Annona muricata, custard apple, i. 69.
Antiquities found in North-America, iii. 123.
Ants, black, ii. 68.
—— red, ii. 70.
Apocynum androsæmifolium, iii. 26.
———————— *cannabinum*, i. 131. ii. 131.
Arctium lappa, burdock, iii. 27.
Ardea Canadensis, ii. 72.
Arum Virginium, Virginian wake-robin, i. 125.
Arundo arenaria, iii. 210.
Asclepias Syriaca, iii. 28.
Asp, Pensylvanian, ii. 125.
Azalea lutea, i. 66.
———— *nudiflora*, white honey-suckle, or May flower, ii. 169.

B.

Badger, i. 189.
Bark-boats, method of making, ii. 298.

Battoes, ii. 242.
Bay St. Paul, in Canada, iii. 200.
Bears carnivorous in North-America, i. 116.
―― plentiful in Canada, iii. 12.
Beavers, ii. 59.
―――― flesh eaten in Canada, iii. 297.
―――― tree, i. 204.
Betula alnus, i. 67. ii. 90.
―― *lenta*, i. 69.
―― *nana (pumila*, Linn.) i. 138.
Bidens bipinnata, i. 171.
Bill of mortality for Philadelphia, i. 57.
Blatta Orientalis, ii. 13, 14.
Blubbers, i. 15.
Blue-bills. See *Anas acuta*.
Blue-bird, ii. 70.
Boats used in Canada, iii. 15.
Bonetos, i. 21.
Bottle-nose, a kind of whale, i. 18.
Bugs, ii. 11.
Bull-frogs. See *Rana boans*.
Bunias cakile, iii. 211.
Burdock. See *Arctium lappa*.
Burlington, the principal town in New-Jersey, ii. 219.

C.

Calabashes, i. 348.
Cancer minutus, i. 13.
Candleberry-tree, i. 192.
Canoes, ii. 241.
Cap aux oyes, iii. 210.
Caprimulgus Europæus, β, whip-poor-Will, ii. 152.
Carabus latus, ii. 68.
Carduus crispus, iii. 294.
Carpinus Betulus, i. 68.
―――― *Ostrya*, i. *ibid*.
Cassia Chamæcrista, i. 120
Castor zibethicus, ii. 57, 285.
Caterpillars, a kind of, ii. 7.
Cattle, wild, i. 207.

Cattle,

INDEX.

Cattle, wild, in the country of the Illinois, iii. 60.
Celtis occidentalis, nettle-tree, i. 69.
Cephalanthus occidentalis, button-wood, *ibid.*
Cercis Canadensis, fallad-tree, *ibid.*
Champlain, lake, ii. 90.
Characters of the French and English women in North-America compared, iii. 55.
———— of the ladies in Canada, iii. 208.
Chenopodium album, i. 118.
——————*anthelminticum*, i. 163.
Chermes alni, i. 154.
Cherry-trees, wild, iii. 160.
Chine, a village in Canada, iii. 305.
Chinquapins, iii. 296.
Cimex lacustris, ii. 126.
—— *lectularius*, ii. 11.
Cicindelæ campestris, varietas, ii. 126.
Civility of the inhabitants of Canada, iii. 135.
Clergy of Canada, iii. 140.
Climate, difference of, between Montreal and Quebec, iii. 152.
Cockroaches, ii. 13.
Cohoes fall, in the river Mohawk, ii. 275.
Collinsonia Canadensis, i. 197.
Coluber constrictor, black snake, ii. 202.
Columba migratoria, ii. 82.
Comarum palustre, i. 138.
Copper, native, from the Upper Lake, iii. 278.
Cornua Ammonis, petrified, iii. 23.
Cornus Florida, dog-wood, i. 66.
Corvus cornix, crow, ii. 66.
Corylus avellana, ii. 90.
Coryphæna Hippurus, i. 19.
Cows in Canada degenerate, iii. 188.
Cranes, American, ii. 72.
———— formerly abundant in America, i. 290.
Cratægus crus galli, i. 66, 115.
———— *tomentosa*, currants, ii. 151.
Crickets, ii. 10.
————field, ii. 69.
Crows, great flights of, ii. 65.

Crystals,

INDEX.

Cryſtals, tranſparent, i. 82.
Culex pipiens, muſquetoes, i. 143.
——— *pulicaris*, ii. 296.
Cunila pulegioides, penny-royal, i. 194.
Cupreſſus thyoides, white cedar, or white juniper, ii. 174.

D.

Dandelion, iii. 13.
Datura ſtramonium, i. 152.
Deal, i. 2.
Decay of the teeth of the Europeans in North-America, i. 360.
Delaware bay, i. 10.
——————river, i. 11.
———————— good water of, i. 47.
———————— convenient for trade, *ibid.*
Delphinus Phocæna, i. 17.
Diet in Canada, iii. 182.
Diospyros Virginiana, perſimon, i. 68, 127, 345.
Dirca paluſtris, mouſe-wood, ii. 148.
Diſtempers common among the Indians, iii. 32.
Dog-fiſh, i. 18.
Dogs trained to draw water from the river, iii. 185.
——— put before ſledges in winter, iii. 186.
Dolphin, or dorado, i. 19.
Draba verna, ii. 91.
Dracontium fœtidum, ii. 90.
Drowned lands, iii. 1.
Dytiſcus piceus, ii. 127.

E.

Elizabeth Town in New-Jerſey, i. 232.
Elymus arenarius, ſea-lime graſs, iii. 210.
Emberiza hyemalis, ii. 51.
Epigæa repens, creeping ground-laurel, ii. 130.
Eſcharæ, i. 13.
Eſquimaux, a nation in the arctic parts of North-America, iii. 233.
——————— arms of, iii. 236.
——————— boats of, iii. 235.

Eſquimaux,

INDEX.

Esquimaux, dress of, iii. 234.
────── words, iii. 239.
Evergreens in North-America, i. 360.
Excrescences on several trees, ii. 22.
Exocoetus volitans, flying fish, i. 20.

F.

Fagus castanea, chestnut-tree, i. 67.
────── *sylvatica,* beech, i. 69.
────── *pumila,* chinquapin, iii. 296.
Fans, made of wild turkeys tails, iii. 66.
Felis lynx, wolf-lynx, ii. 200.
Fever and ague, i. 364.
Fish, caught by a peculiar method at Trois Rivieres, iii. 92.
────── flying, i. 20.
Fleas, original in America, ii. 9.
Food of the Indians, ii. 95.
Formica nigra, ii. 68.
Fort St. Frederic, or Crownpoint, iii. 4, 34.
────── John, in Canada, iii. 45.
Foxes, grey, i. 282.
──────red, i. 283.
Fraxinus excelsior, ash, i. 68.
Fucus natans, sea-weed, i. 12.

G.

Galium tinctorium, iii. 14.
Gentiana lutea, i. 138.
────── *saponaria,* iii. 294.
German-town in Pensylvania, i. 89
Giants pots, i. 121.
Ginseng, iii. 114.
Gleditsia triacanthos, honey-locust-tree, i. 69.
Glycine Apios, ii. 96.
Gnaphalium margaritaceum, i. 130.
Goods that have a run among the Indians, iii. 266.
────── given in exchange by the Indians, iii. 274.
Gourds, i. 347.
Gracula quiscula, the purple daw, ii. 76.

Grass-

INDEX.

Grafs-worms, ii. 76.
Ground-hog. *See* Badger.
Gryllus campestris, ii. 10, 69.
—— *domesticus*, ii. 10.
Gulls, common, i. 23.
Gypfum, fibrous, iii. 229.
Gyrinus natator (Americanus), ii. 139.

H.

Hamamelis Virginica, i. 68.
Hares in Canada, iii. 59.
Hatchets of the Indians, ii. 37.
Hedera helix, ivy, i. 141.
Helleborus trifolius, iii. 160.
Hinds, tamed in North-America, ii. 197.
Hinlopen cape, in Penfylvania, i. 10.
Hirundo pelasgia, chimney-fwallow, ii. 146.
—— *purpurea*, purple-martin, ii. 147.
—— *riparia*, fand-martin, or ground-fwallow, ii. 147.
—— *rustica*, barn-fwallow, ii. 140.
Hopnifs. See *Glycine Apios*.
Horfes in Canada ftrong, iii. 187.
Humming bird, i. 210.
Hurons, an Indian nation, iii. 178.

I.

Jerfey pine, i. 334.
Ilex aquifolium, holly, i. 351, 360.
Impoffibility of eftablifhing filk manufactures, and making wine, in North-America, i. 123, 125.
Indians, livelihood of the, ii. 113.
—— religion of the, ii. 117.
Inhabitants of Canada, iii. 8.
Inftances of great fertility among the inhabitants of North-America, ii. 4.
Intenfenefs of the froft in America, ii. 49.
Iron-works at Trois Rivieres, iii. 87.
Juglans alba, hiccory, i. 66.
—— *nigra*, i. 67.
—— *baccata?* butternut-tree, i. 69.

Juniperus

INDEX.

Juniperus Virginiana, the red cedar, or red juniper, ii. 180.

K.

Kalmia latifolia, i. 68, 336.
———— *angustifolia*, ii. 215.
Katnifs. See *Sagittaria sagittifolia*.
Kettles of the Indians, ii. 41.
Kitchen-herbs of Canada, iii. 129.
Knives of the Indians, ii. 39.

L.

Lac St. Pierre, iii. 83.
Land-birds seen at sea, i. 24
Larus canus, i. 23.
Laurus æstivalis, spice-wood, i. 68.
———— *sassafras*, i. 68, 146, 340.
Lead-veins near Bay St. Paul, iii. 212.
Leontodon taraxacum, iii. 13.
Lepas anatifera, i. 16.
Licences for marrying in America, in the gift of the governors, ii. 25.
Lichen rangiferinus, iii. 137.
Ligustrum vulgare, privet, i. 86, 165.
Lime-flates, black, iii. 243.
Lime-stone, pale grey, i. 84.
Linnæa borealis, i. 138.
Liriodendron tulipifera, i. 66, 202.
Liquidambar styraciflua, i. 67, 161.
Locusts, which destroy the young branches of trees, ii. 6.
Log-worms, i. 2.
Long island, ii. 226.
Loxia Cardinalis, ii. 71.
Lupinus perennis, ii. 155.
Lynxes in America. See *Felis lynx*.

M.

Magnolia glauca, beaver-tree, i. 69, 204.
Maize-thieves, description of, ii. 74.
———————— natural history of, ii. 76.

INDEX.

Maize-thieves, proscribed in America, ii. 78.
——————— white backed, ii. 274.
Maple, red, i. 167.
Marangoins, a kind of gnats, iii. 47.
Marble, white with blueish grey spots, i. 83.
Marmor rude. See Lime-stone.
Mechanicks, few in Canada, iii. 59.
Medusa aurita, i. 15.
Meloe majalis, ii. 105.
—— *proscarabæus*, ii. 157.
Mickmacks, an Indian Nation, iii. 180.
Mink, ii. 61.
Mocking bird, i. 217.
Moles, a kind of, i. 191.
——— subterraneous walks of, i. 190.
Montmorenci waterfall, iii. 227.
Montreal, a great town in Canada, iii. 71.
——————— account of the climate of, iii. 75.
——————— churches and convents of, iii. 72.
——————— hospital of, iii. 74.
Moose-deer, i. 296.
——————— nothing but an elk, iii. 204.
Morus rubra, i. 68.
Motacilla sialis, blue bird, ii. 70.
Moths abundant in the clothes and furs, ii. 8.
Mountain flax, i. 303.
Muscovy glass, i. 84.
Musk rats, ii. 56.
——————— carnivorous, ii. 285.
Musquitoes, i. 113.
Myrica cerifera, candleberry-tree, i. 192.
———— *gale*, i. 138.
Mytilus anatinus, muscle shells, ii. 80, 114.

N.

Natural history promoted in Canada, iii. 5.
Negroe slaves in North-America, i. 396.
——————— know a kind of poison, i. 397.
New Bristol, i. 219.
—— Brunswick, i. 229.

New

INDEX.

Newcaftle, a town in Penfylvania, i. 26.
────────── founded by the Dutch, i. 26.
New-York, i. 247.
────────── affembly of deputies, i. 259.
────────── houfes of, i. 249.
────────── public buildings, i, 250.
────────── port, i. 252.
────────── trade, i. 253.
Nicholfon fort, near Canada, ii. 293.
Noxious infects in America, ii. 6.
Nyſſa aquatica, Tupelotree, ii. 67.

O.

Oenothera biennis, iii. 294.
Oriolus phoeniceus, ii. 79.
Orleans, Ifle of, in the river St. Lawrence, iii. 194.
Orontium aquaticum, ii. 101.
Oxalis corniculata, i. 201.

P.

Panax quinquefolium, iii. 114.
Paper-currency of Canada, iii. 68.
Papilio antiopa, ii. 105.
────────── *euphroſyne,* ii. *ibid.*
Parfneps, iii. 67.
Partridges, American, ii. 51.
────────── white. See Ptarmigans.
Parus major, i. 24.
Peafe, deftroyed by an infect, i. 173.
Pectinites, iii. 22.
Penn's Neck. in New-Jerfey, ii. 17.
Petite Riviere, iii. 221.
Petrel, i. 23.
Phaëton æthereus, i. *ibid.*
Philadelphia, capital of Penfylvania, i. 31.
────────── by whom, and when built, i. 32.
────────── houfes of, i. 34.
────────── public buildings, i. 36.
────────── regularity, and beauty of its ftreets, i. 33.

INDEX.

Philadelphia, temperature of its climate, i. 46.
────────── trade of, i. 49.
Phytolacca decandra, American nightshade, i. 95, 196.
Picus auratus, ii. 86.
────── *carolinus*, ii. ibid.
────── *erythrocephalus*, ii. ibid.
────── *pileatus*, ii. ibid.
────── *principalis*, ii. 85.
────── *pubescens*, ii. 87.
────── *varius*, ii. ibid.
────── *villosus*, ii. 86.
Pierre à Calumet, iii. 230.
Pigeons, wild, ii. 82.
Pinus abies, the pine, i. 360.
────── *sylvestris*, the fir, i. ibid.
────── *tæda*, i. 69.
────── *Americana*, i. ibid.
Plantago major, i. 118.
────────── *maritima*, iii. 211.
Platanus occidentalis, i. 62.
Pleurisy, i. 376.
Poa angustifolia, iii. 156.
────── *capillaris*, iii. 66.
Poke. See *Phytolacca*.
Polecat, American, i. 273.
Polytrichum commune, i. 184.
Pontederia cordata, iii. 260.
Porpesse, i. 16.
Portuguese, or Spanish man of war, a species of blubber, i. 15.
Portulaca oleracea, purslane, ii. 284.
Potentilla fruticosa, i. 138.
Prairie de Magdelène, a small village in Canada, iii. 52.
Preferableness of Old Sweden to New Sweden, ii. 188.
Prinos verticillatus, i. 67.
Probability of Europeans being in North-America long before Columbus's discovery, ii. 31.
Procellaria pelagica, i. 22.

Prunella

Procellaria puffinus, i. 23.
Prunella vulgaris, iii. 294.
Prunus domestica, i. 67.
———— *spinosa*, i. 68.
———— *Virginiana*, i. 67.
Ptarmigans, iii. 58.
Pyrites, cubic, i. 82.
Pyrus coronaria, crabtree, i. 68. ii. 166.

Q.

Quebec, the chief city in Canada, iii. 97.
———— the palace of, iii. 99.
———— other public buildings, iii. 100.
———— climate of, iii. 246.
Quercus alba, i. 65.
———— *Hispanica*, i. 66.
———— *phellos*, ibid.
———— *prinos*, ibid.
———— *rubra*, ibid.
———————— varietas, i. 68.

R.

Raccoon, i. 97, ii. 63.
Rana boans, bullfrog, ii. 170.
———— *ocellata*, ii. 88.
Rapaapo, a village in New-Jersey, ii. 168.
Rats, not natives of America, ii. 47.
Rattle-snake, found no further north than fort St. Frederick, iii. 48.
Reasons for supposing part of North-America was formerly under water, i. 132, i. 199.
Redbird, ii. 71.
Rein-deer moss, iii. 137.
Remarks upon the climate of North-America, i. 106.
Rhus glabra, sumack, i. 75, 66.
———— *radicans*, i. 67, 177.
———— *vernix*, poison tree, .. 77, 68.
Ribes nigrum, i. 68.
Robinia pseudacacia, locust-tree, i. 69.
Robin-red-breast, American. See *Turdus migratorius*.

Rockstones of various sorts, near Fort St. Frederic, iii. 20.
Rubus occidentalis, i. 66.
Rudbeckia triloba, iii. 294.

S.

Sagittaria sagittifolia, ii. 97.
Salem, a little town in New-Jersey, ii. 164.
Sambucus occidentalis, s. Canadensis, i. 66. ii. 283.
Sands of several sorts, near Lake Champlain, iii. 24.
Sanguinaria Canadensis, ii. 140.
Saratoga, an English fort towards Canada, ii. 289.
Sarothra gentianoides, i. 126.
Scarabæus, ii. 68.
——— *carolinus?* ii. 125.
Scirpus pallustris, iii. 83.
Scomber pelamys, boneto, i. 21.
——— *thynnus*, tunny, i. 19.
Sea hen, i. 24.
Sea weeds, i. 12.
Servants, different kinds of, i. 387.
Shear water, i. 23.
Ships, annually entered into, and sailed from Philadelphia, i. 53.
Sison Canadense, iii. 27.
Skeleton found in Canada, supposed to be of an elephant, iii. 12.
Skunk, or American pole-cat, i. 273.
Smilax laurifolia, i. 68. ii. 185.
Snake, black, ii. 202.
Snow-bird, ii. 51, 81.
Soap-stone, i. 300.
Soeurs de Congregation, iii. 304.
Soldiers advantageously provided for in Canada, iii. 16.
Sorbus aucuparia, iii. 151.
Spartium scoparium, i. 287.
Squashes, i. 348.
Squirrels, flying, i. 320.
——— grey, i. 310.
——— ground, i. 322.

State,

INDEX.

State, former, of New-Sweden, ii. 106.
—— of the American Indians before the arrival of the Europeans, ii. 36.
Sterna hirundo, i. 23.
Sturgeons, ii. 229, 278.
Sulphureous springs near Bay St. Paul, iii. 215.
Swallow, barn or house, ii. 14.
—— chimney, ii. 146.
—— ground, or sand martin, ii. 147.
—— seen at sea, i. 24.

T.

Tawho, or Tawhim, ii. 98.
Tawkee. See *Orontium*.
Terns, i. 23.
Tetrao lagopus, Ptarmigans, iii. 58.
Thuja occidentalis, iii. 170.
Tilia Americana, lime-tree, i. 59.
Tisavojaune rouge, iii. 14.
Titmouse, great, i. 24.
Tobacco pipes, Indian, ii. 42.
Travado. ii. 214.
Trees, which resist putrefaction less than others, ii. 19.
Trientalis europæa, i. 138.
Triglochin maritimum, i. 138.
Trochilus colubris, i. 210.
Trois Rivieres, a town in Canada, iii. 85.
Tropic bird, i. 23.
Turdus migratorius, ii. 90.
—— *polyglottos*, ii. 217.
Turtle, i. 22.
Typha latifolia, ii. 132. iii. 218.

U.

Vaccinium, a species of, i. 66.
———— another species, *ibid*.
———— *hispidulum*, ii. 79.
Veratrum album, ii. 91.
Verbascum thapsus, i. 128.
Verbena officinalis, i. 119.

Viola

Viola Canadensis, iii. 294.
Viscum album, i. 360.
────── *filamentosum*, i. 286.
Vitis labrusca, i. 66.
────── *vulpina*, ibid.
Viverra putorius, skunk, i. 273.
Ulmus Americana, i. 67. ii. 298.
Ursus Meles, badger, i. 189.

W.

Wampum, ii. 261. iii. 273.
Wasp-nests, curious, ii. 137.
Water, bad at Albany, ii. 253
Watering of meadows in Pensylvania, i. 308.
Water-melons, iii. 261.
Waves, bigness of, in the Bay of Biscay, I. 3.
Whip-poor-Will, ii. 151.
Whortle-berries, American, ii. 80.
Wilmington, a little town in Pensylvania, i. 156.
Winds, changeable about the Azores, i. 5.
Wolves in America, i. 285.
Women in Canada, dress of, iii. 81.
Wood of different sorts, for joiners work, ii. 21.
Woodbridge, a small village in New-Jersey, ii. 232.
Woodlice, ii. 16. 303.
Woodpeckers, Carolina, ii. 86.
────────── crested, *ibid*.
────────── gold winged, *ibid*.
────────── king of the, ii. 85.
────────── least spotted, ii. 87.
────────── lesser spotted, yellow bellied, *ibid*.
────────── red headed, ii. 86.
────────── seen at sea. i. 25.
────────── spotted hairy, ii. 86.

Z.

Ziz ania quatica. iii. 32, 54.

F I N I S.

www.ingramcontent.com/pod-product-compliance
Lightning Source LLC
Chambersburg PA
CBHW031859220426
43663CB00006B/694